ARAB WATER SECURITY
Threats and Opportunities in the Gulf States

Access to a secure supply of freshwater is critically important in the Arab Gulf states. With a growing dependence on large desalination plants, the threats to water security, and, in turn, food, energy, and national security, are a real and pressing concern.

This book explores the national security implications of the Arab Gulf states' reliance on desalination plants, and their related infrastructure. It provides the first systematic and comprehensive discussion of current and future threats to the supply of freshwater from a desalination plant, including actual and virtual attacks by terrorists, mechanical failure, contamination, sabotage by aggrieved workers, and attacks relating to regional conflicts, as well as their vulnerability to natural disasters. It also provides a detailed analysis of the effects of a potential disruption to the water supply, and proposes possible measures, both political and technological, that can be used to increase resilience to these threats.

Arab Water Security is a valuable reference for researchers and graduate students, as well as for policymakers and professionals, interested in water security, natural resources, and environmental terrorism.

DR. HUSSEIN A. AMERY is an Associate Professor in the Division of Liberal Arts and International Studies at the Colorado School of Mines, where he teaches courses on the political economy of resources, the Middle East, and on water politics and policy. His academic expertise is in human and environmental security, transboundary water conflicts, and in identifying and analyzing threats to critical infrastructure in the Arab Gulf states and the wider Middle East. He was selected as Fellow by the International Water Association in 2005, and has served as consultant on desalination and other water issues to the American and Canadian governments, as well as engineering firms.

ARAB WATER SECURITY

Threats and Opportunities in the Gulf States

HUSSEIN A. AMERY

*Division of Liberal Arts and International Studies,
Colorado School of Mines*

CAMBRIDGE
UNIVERSITY PRESS

CAMBRIDGE
UNIVERSITY PRESS

University Printing House, Cambridge CB2 8BS, United Kingdom

Cambridge University Press is part of the University of Cambridge.

It furthers the University's mission by disseminating knowledge in the pursuit of education, learning and research at the highest international levels of excellence.

www.cambridge.org
Information on this title: www.cambridge.org/9781107042292

© Hussein A. Amery 2015

First published 2015

A catalogue record for this publication is available from the British Library

Library of Congress Cataloguing in Publication data
Amery, Hussein A., 1958– author.
Arab Water Security: Threats and Opportunities in the Gulf States / Hussein A. Amery,
Division of Liberal Arts and International Studies, Colorado School of Mines.
pages   cm
ISBN 978-1-107-04229-2 (Hardback)
1. Water security–Persian Gulf States.   2. Food security–Persian Gulf States.   I. Title.
HD1698.5.A36 2015
333.91009536–dc23   2014050245

ISBN 978-1-107-04229-2 Hardback

To the memory of my late mother, Helwi Najeeb Kassim AbuGhoush.
Your grace, compassion and natural intellect will always inspire me.

To my wife, Maha,
and children Hisham, Laila and Iman.

Contents

List of figures

List of tables

Preface

In the arid desert landscape of the Arab Gulf states, water is everywhere, and there is plenty of it to drink. This is in sharp contrast to the "Rhyme of the Ancient Mariner" that spoke of "water, water everywhere, nor any drop to drink" to describe the mariners of the Middle Ages who were adrift in the sea with nothing to drink (O'Brien, 2006; Hurlimann, 2006). The hydrocarbon wealth discovered in the Arabian Peninsula some eight decades ago eventually allowed the countries that emerged there to desalinate large quantities of seawater to help sustain the populations, their globalized economies, and modern infrastructures with all the amenities and water-intensive lifestyles that are associated with modern living. Today, Gulf governments often decorate major street intersections with water fountains, and line highways with green belts that typically have grassy landscape and palm trees that are irrigated regularly. Also, natives and visitors enjoy lots of swimming pools, lush green grass in private and public spaces, and delight in large water theme parks, and in indoor, multi-story high down-hill ski slopes. When you combine such ubiquitous amenities with free-of-charge freshwater and with household faucets that never desist, one then understands why so many residents of the Arab Gulf states are unaware of the severe water shortages that are pervasive in their region. Although water security is about managing both supply and demand, the young Gulf states have been focusing on water supply and have not paid sufficient attention to limiting demand through efficiency improvements. The Gulf states of today are less than 50 years old; the Kingdom of Saudi Arabia is much older having gained its independence in the early 1920s. Given this, the six Gulf countries have been undergoing an all-encompassing process of nation building.

Water security is closely related to natural and contrived water scarcity. The Gulf states' experiences are somewhat unique because the geographic space that they occupy lacks any perennial river system, and, decades ago, Gulf governments realized the need to supplement their rapidly dwindling groundwater and chose the desalination technology. The following conceptual narrative of water development in the Gulf states is an adaptation and expansion of the literature on the politics of water development and management of river systems (Molden *et al.*, 2001; Keller *et al.*, 1998; WWDR, 2012, and FAO, 2008).

Water development in the Gulf states went through phases of exploitation, utilization, augmentation, and conservation. Until the early twentieth century, the Gulf region was lightly populated and poorly developed. At that time, the people used mostly non-mechanical tools to draw water from shallow aquifers which satisfied their basic needs. They eventually built distribution systems, as well as storage and treatment facilities. As the demand for water increased and modern technology became available, the people started to drill deeper wells using fuel-powered pumps, transport water to more distant places, enhance water productivity, import some water-intensive food crops, and to eventually build desalination plants. As economic maturity started to set in, countries crafted long-term plans to ensure the environmental, economic, and social sustainability of what had been erected in previous decades. This usually includes measures to mitigate water pollution, wasteful use, and reallocation to more efficient uses. The Gulf countries have been in water supply phases for too long, and are starting to take hesitant steps towards water demand management.

Deliberate or accidental technological failures, as well as cyber-attacks that result in prolonged and catastrophic disruption of supply, could have significant ramifications for political stability. Water-supply disruptions could be caused by factors such as natural hazards, sectarian discord that permeate Gulf societies, a few low-skilled guest workers acting on their grievances, and by acts of terrorism that may spill-over from near-by countries through infiltrations of the porous borders or through home-grown terrorists who may conduct collective or "lone-wolf" strikes on behalf of a foreign group or ideology.[1]

While the Gulf states have generally been politically stable, a simmering unease dwells beneath this apparent tranquillity.[2] Variable socio-economic and political crosswinds continue to blow over the Middle East and the Arabian Peninsula such as the Arab Spring, high unemployment rate among nationals, as well as medium to high-intensity sectarian wars in Yemen, Iraq, and Syria. Would one or a few of these factors be strong enough to force rulers to embrace the necessary reforms? If that happens, would the transition to the new political reality be smooth and peaceful? Are current institutions sufficiently dynamic and robust to respond to peoples' wishes and initiate gradual political reform? Is the

---

[1] See Asal *et al.* (2013) for details on the threat posed by lone-actor terrorists.
[2] For details about the multitude of threats that could undermine the prevailing political order in the Arab Gulf States, see Davidson (2013).

monarchical political system sufficiently resilient to make tough decisions such as graduated lifting of water subsidies?

Prevailing social and political stability often lulls governments into assuming that their countries don't have looming serious security risks. Contrary to this belief, the approach followed in this volume is inspired by the well-known insights of the ancient Chinese military strategist Sun Tzu who advised in his famous book *The Art of War* that countries should not assume that the enemy will not come, but should be prepared for his coming; they should not presume he will not attack, but instead make their own positions unassailable. The governments of the Gulf states, especially after the 2001 attacks in the United States, have increased the size of and capabilities of their security services, hardened their critical infrastructure, and expanded their technological defences. This book examines the security and vulnerability of desalination technology, and whether Gulf states have taken steps to "harden" their culturally and linguistically diverse societies, including members of the marginalized classes.

The genesis of this book project was started over ten years ago. I was fortunate to have had the opportunity to spend a year teaching and researching at the Petroleum Institute in Abu Dhabi, and later spent shorter research periods of time in different countries in the Gulf. The ideas in this book benefitted from observations, interviews with a large number of people ranging from natives ("locals") and civil servants, to water professionals, academics, and with a large number of guest workers who hailed from various Arab countries, Africa, southeast and south Asia, Africa, and from Western Europe and North America. The constricting political and security culture makes the vast majority of foreigners fear for their jobs which, if lost, would result in them being deported from their host country. This deters the majority, especially labourers and maids, from publicly expressing their grievances and from speaking on the record about their life in a Gulf country. Arguments and ideas in this book were also improved by feedback received from experts in the field at different specialized conferences and symposia that in cities as varied as Bangkok, Muscat, Abu Dhabi, Calgary, Tampa, and Denver. Finally, I owe a debt of gratitude to the many graduate students at the Colorado School of Mines who, over the years, have helped me find qualitative and quantitative data that I needed for this project.

Ultimately, it is my hope that this volume will contribute to raising the profile of water security in the Gulf region (and the wider Middle East), to highlighting the potential security threats that lurk over the political and social horizons, and to inspiring threat-mitigation measures that would improve the security environment and the quality of life of all the stakeholders. This is perhaps a tall order. Inaction, however, could aggravate threats to water infrastructure and increase the likelihood of a high-impact water-disruption event.

1

Rethinking water and food security in the Arab Gulf states

1.1 Introduction

A scientific journal reported the discovery of seven-million-year-old footprints of elephants, the world's oldest elephant tracks, in the United Arab Emirates (UAE). Paleontologists say the area had much more water, vegetation, and animal life, and its biodiversity resembled what was present in wet parts of Africa and Europe. "The region then was home to a great diversity of animals, including elephants, hippopotamuses, antelopes, giraffes, pigs, monkeys, rodents, small and large carnivores, ostriches, turtles, crocodiles, and fish. These were sustained by a very large river flowing slowly through the area, along which flourished vegetation, including large trees. The animals resembled those from Africa during the same time, though there are also similarities with Asian and European species of that period" (Livescience.com, 2012). The image of a tropical paradise teeming with biotic life is today a polar opposite of what it used to be.

Social stability and economic prosperity rest on regular access to sufficient amounts of potable water. Many areas of the world may have hit "peak water," which explains the growing talk about a water crisis, and why this resource, long taken for granted, is now being called the "new oil" or "blue gold." Throughout history, humans have always been well tuned to nature's rhythms that helped them harness and secure the resources they needed for their survival. They devised new ways for identifying water sources, and for harvesting, transporting, and storing water so to meet their then-simple and basic needs. How will technological advancements affect water security of future generations?

Water security is a complex, multi-dimensional concept. Water insecurity is a relative concept because, at one level, it is an imbalance between water "supply" and "demand" and is affected by spatial, temporal, and economic conditions (Jägerskog *et al.*, 2014; Swain, 2012). It is also a dynamic process because it is aggravated by higher human demands, varying supplies, degraded quality of the resource, and by poor governance and inadequate policy response. A recent report by the World Economic Forum (2014) ranked the water crisis as the third most important challenge facing the world. This measure by prominent political and economic leaders from around the world helps focus the attention of governments, businesses, and civil societies on this issue that has been rising in importance. Water security is affected by physical availability and technological ability to produce potable water, and by a government's ability to develop institutions and build the infrastructure necessary to ensure a reliable supply of water.

People and governments increasingly view water as a resource of strategic importance, one that affects human and national security. Fearing an interruption of the supply of critical resources appears to be a common human concern that is unrelated to geography or culture. A recent survey on "British attitudes towards the UK's international priorities" revealed that for most people (53 percent) in the country the biggest future threats to "the British way of life" were terrorism, followed by interruptions to energy supply (37 percent), and "long-term scarcity of essential natural resources, such as water, food and land" (30 percent). Climate change (18 percent) was the fourth and final item on the list of apprehensions (Chatham House, 2011). In another poll, most (48 percent) opinion leaders and decision makers in Britain said that the main focus of their country's foreign policy should be to ensure "the continued supply of vital resources, such as oil, gas, food and water [tied with terrorism]" (Evans 2011). The British fear appears irrational when you consider that the average annual precipitation for the United Kingdom is 1,222 mm (World Bank Data, n.d.). On the other hand, the Gulf states have an average annual precipitation that is well under 125 mm (World Bank Data, n.d.), are void of perennial rivers, and, for the last few decades, have experienced much higher levels of economic and population growth than many countries, including Britain. This, along with their rapidly improving quality of life, have resulted in an astronomical rise in their total and per capita water consumption. As devoid as they are of water, their significant endowments in hydrocarbon wealth has made it possible for them to overcome their physical scarcity of water and food.

The Arab world, from Iraq to Egypt and all the way to Morocco, is one of the most arid regions on the planet. Within this large cultural region is the geopolitical sub-region known as the Gulf Cooperation Council (GCC), which includes Saudi Arabia, UAE, Kuwait, Qatar, Bahrain, and Oman. Although Yemen and Iraq are non-members, the turmoil and wars they have long experienced pose serious security challenges to the GCC. It has been trying to contain the socio-political spill-over effects of these and other wars on their countries. Since 1980, Iraq and Yemen have experienced wars and violent insurgencies. Iraq, for example, invaded Iran in 1980, triggering an eight-year war, and then invaded Kuwait in 1990.

Kuwait's desalination plants were thought to be within the range of Iraqi and Iranian missile placements and therefore "easily targetable" (Cordesman, 1997, p. 58). The United States' 2003 war on Iraq led to the collapse of the nation state and its institutions, which in turn fueled waves of insurgency and terrorism that continued well into 2014. When the once-divided Yemen was united in 1990, many people in the northern and southern parts of the country did not buy into the social and political integration, leaving the country in the throes of perpetual turmoil. High unemployment and political instability drive many Yemenis to seek a better life in the wealthier Gulf states, and some profit from illegal activities such as smuggling people, weapons, and drugs into Saudi Arabia. Finally, Al Qaeda in the Arabian Peninsula is based in southern Yemen.

The natural environment is the original culprit in the hydrological conditions that Arabs contend with. The Arab world is located in a generally arid to very arid region where environmental conditions have gradually worsened for the people. The renewable water resources available in 1950 were over 4,000 m^3/capita per year, declined to 1,312 m^3/capita per year in 1995, and slipped to 1,233 m^3/capita per year in 1998; they are projected to reach 547 m^3/capita per year in 2050 (Arab Water Council, 2009). Yemen, one of the most water-deficient countries in the world, has an annual per capita water availability of only 125 m^3, compared to the global average of 2,500 m^3 (WWAP, 2012). The freshwater that is available for the people of Yemen or for the GCC countries is significantly lower than the global average, which underscores the severity of the situation in this region. Scholars classify a country as experiencing "water stress" when its annual renewable water supplies fall below 1,700 m^3 per person, "water scarcity" when they reach 1,000 m^3 per person, and as having "absolute scarcity" when they dip below 500 m^3 per person (WWAP, 2012). All of the Gulf states suffer from an acute case of absolute water scarcity.

Over the past five decades, these countries have experienced a dramatic increase in population and in the quality of life that strained their natural water supplies so much that they looked for alternative sources that would supplement their aquifers; they chose to desalinate seawater to meet their domestic freshwater needs. Furthermore, since the 1960s, the Arab Gulf states have experienced dramatic increases in population sizes due to high natural growth rates and, more importantly, due to the very high influx of foreign workers. This, together with the fast pace of modernization and urbanization have rapidly inflated the size of primate cities like Dubai (1.9 million), Riyadh (5.5 million), Jeddah (3.6 million), and Kuwait city (2.4 million) (CIA, 2011). Current research predicts that the availability of renewable freshwater resources will continue to decrease with changes in precipitation and recharge rates of groundwater resources. By 2050, renewable water resources in many Middle Eastern countries – including those in the GCC – will decrease between 25 and 40 percent (FutureWater, 2011).

Among the GCC countries, Saudi Arabia and the UAE have the largest population sizes, and therefore experienced the largest net increases. Yet, even though Bahrain only has 1.3 million people (2013), the country's population has increased by over 700 percent since 1960 (Table 1.1).

With the significant increase in population and the demand for water, the natural water resources of the Arabian peninsula have been stretched and are not enough to support the population. For all Gulf countries, the total renewable water resources per capita have decreased dramatically (Table 1.2). Since the 1960s, the renewable water resources per capita in Oman, Saudi Arabia, Bahrain, and Kuwait have decreased by 82, 84, 86, and 89 percent, respectively. The biggest decreases were in Qatar and the UAE where population growth skyrocketed and their renewable water resources per capita have fallen by 97 percent and over 98 percent, respectively.

Table 1.1 *Population size in 1960 and 2013, and the percent increase*

Country	Total population, 1960	Total population, 2013	Percentage population increase, 1960–2013
Bahrain	162,501	1,332,171	719.8
Kuwait	261,994	3,368,572	1,185.7
Oman	551,737	3,632,444	558.4
Qatar	47,316	2,168,673	4,483.8
Saudi Arabia	4,072,110	28,828,870	608.0
UAE	89,608	9,346,129	10,330.0
Yemen*	5,099,785	24,407,381	378.6
Tunisia*	4,220,701	10,886,500	157.9
Hungary*	9,983,967	9,897,247	−0.8
Canada*	17,909,009	35,158,304	96.0

* These countries are intended to serve as a rough reference point.
Source: Statistics calculated from World Bank Databank, databank.worldbank.org.

Table 1.2 *Total renewable water resources per capita (m³/inhabitant/year)*

	1962	1992	2012
Bahrain	670.5	221.8	88.01
Kuwait	59.7	10.6	6.15
Oman	2,418	863.1	422.50
Qatar	1,036	137.1	28.28
Saudi Arabia	552.2	165.2	84.84
UAE	1,376	99.0	16.29
Yemen	398	201.7	88.04

Source: AQUASTAT, United Nations, http://www.fao.org/nr/water/aquastat/dbases/index.stm.

Table 1.3 *Average annual precipitation levels in the Gulf states and Yemen*

Country	Bahrain	Kuwait	Oman	Qatar	Saudi Arabia	UAE	Yemen
Precipitation (mm/yr)	83	121	125	74	59	78	167

Source: AQUASTAT, United Nations, www.fao.org/nr/water/aquastat/.

The Arab Gulf states share largely similar climatic conditions but there are some important variations between them. Except for a few areas, most of the peninsula is either classified as desert or semiarid mountains. Although the Oman Mountains and Asir Mountains in south and western parts of the Arabian Peninsula enjoy higher rates of precipitation and runoff, there are no perennial rivers or lakes in any of the GCC countries. Rain generally occurs in the winter months, is unpredictable, and often results in flash flooding, temporarily filling wadis. Topographically, much of the region is flat. Combined with high evaporation rates, the topography makes it difficult to harvest rainwater and little of it recharges groundwater before it is evaporated (Al-Rashed and Sherif, 2000). Average annual rainfall (Table 1.3) can be misleading because of the intense evaporation and the significant variation in precipitation between the hyper-arid deserts and the mountainous areas which receive much higher levels of rainfall.

The sporadic and meager surface water resources of the GCC states are minor compared to the large but diminishing fresh groundwater resources throughout the region. Alluvial groundwater aquifers tend to be shallow, recharged by rainwater, and their water of better quality. Non-renewable deep aquifers are not fed by infiltrating rainwater. Groundwater resources in the deep aquifers are estimated at 2,330 billion cubic meters and over 30 percent of the groundwater reserves are located in the Wasia-Biyadh aquifer (one of the largest in Saudi Arabia). Depending on the geological composition, the quality of water varies significantly from one aquifer to another. Overuse due to irrigation has affected the quality and productivity of aquifers. In areas along the coast of GCC countries, saltwater intrusion into freshwater aquifers is a problem (Al-Rashed and Sherif, 2000; Al-Hajri and Al-Misned, 1994).

The smallest of the countries, Bahrain, has an arid to hyper-arid environment with high temperatures, erratic rainfall, and high evaporation rates. Rain only supports the most drought-resistant vegetation. The principle source of water is from the Dammam Aquifer, only a small part of the larger Eastern Arabian Aquifer. However, the aquifer has suffered severe degradation and salinization from multiple sources: (1) brackish-water up-flow from underlying water zones below the aquifer, (2) seawater intrusion, (3) intrusion from saline aquifers, and

(4) return flow from irrigation (FAO, 2008; Musayab, 1988). Over-exploitation of groundwater has also damaged what few wetlands existed and has resulted in the drying of all natural springs (UNDP, 2013).

Most of the soil in the Gulf states does not retain moisture and is made up of many "hard pans," referred to by locals as "gutch," which prevent water from infiltrating into aquifers. This, high evaporation rates, and the over-drafting of aquifers for irrigation have been depleting these mostly non-renewable resources and degrading their quality. The meager rainfall is unpredictable and insufficient to support rain-fed irrigation, but it is the main source of recharge for the few renewable aquifers in the Gulf. Freshwater aquifers in the Gulf states usually lie above saline groundwater and their over-use has caused saline water to up-flow into the aquifer (Lloyd *et al.*, 1987). This mis-management has caused a lowering of the water table and deeper, brackish water to up-flow into freshwater sources. Many of the aquifers are composed from limestone, which has led to severe seawater intrusion and increased salinity of groundwater (AQUASTAT, 2009).

There are groundwater resources in the Bajada region. These aquifers contain water from alluvial fans along the base of the Oman and Ras Al Khaymah mountains. Dams have been built in areas where water infiltrates through permeable streambeds, hence replenishing groundwater (Murad *et al.*, 2007).

While many of the GCC countries are similar in climate and have few options concerning water resources, both Oman and Saudi Arabia are much more environmentally diverse and have more natural freshwater resources than other Gulf states. Oman, for example, is divided into three regions. The coastal plains consist of vital agricultural areas and are hot and humid throughout the year. In the southernmost reaches, monsoons occur during the summer. Even though there are wadis with intermittent surface runoff, internal groundwater is the main reliable source of renewable freshwater. There are several important aquifers in the northern and southern areas and the soil easily allows precipitation to infiltrate into the groundwater. Some of these aquifers are a part of the larger system that extends throughout the peninsula. In addition to these, most other sources of groundwater are brackish to saline. There are also large amounts of freshwater reserves in aquifers that were filled a long time ago when climate conditions were different. These non-renewable resources have a very low recharge rate and have been designated by the government as a reserve for future use (World Bank, 1988), and are being increasingly treated as reserve for future use

Saudi Arabia also has significant water resources and varied climates. The Western Mountains, or "Arab Shield," consists of high peaks, deep valleys, and enjoys the heaviest rainfall in the region. East of these mountains are the central hills and further east of that is a sandy, hot, desert region. The government of Saudi

Arabia has constructed dams in different parts of the country to trap water from the short-lived flash floods, which is used for groundwater recharge, and irrigation. The quality of groundwater varies from area to area but most of it is classified as brackish (Gutub *et al.*, 2013).

Climate change exacerbates prevailing hydrological stress. A study commissioned by the World Bank reviewed the results of nine global climate-change models and reported that the gross recharge between 2010 and 2050 is expected to drop sharply in "almost all" the Middle East and North American (MENA) countries, where the largest declines for that period are expected in Oman, UAE, and Saudi Arabia. The models reveal that the largest decreases in internally and externally renewable water resources in the Gulf will be in Oman (–46 percent) and Saudi Arabia (–36 percent) (Immerzeel *et al.*, 2011, p. 57).

Freshwater, already a scarce resource, is becoming harder to find due to population pressure, mismanagement, and climate change. Desalinated water is critical to the well-being of people of the Gulf and to the modern economies that they have come to depend on. Enduring sustained water-supply disruptions could have serious ramifications on the social and political stability of a country. Therefore, a better understanding of threats to water supplies would gauge the social resilience of affected countries, provide useful information to the business sector, and deliver respective government agencies an early warning, encouraging them to consider preventative or mitigating measures that would ensure water security for all.

Water security has been a dominant concern in certain transboundary water negotiations (Mekonnen, 2010), and has been adopted by major international aid agencies such as those of the German and American governments (BMZ, 2010; USAID, 2014). It has also been linked to economic growth and human development (Liu *et al*, 2007), and associated with sustainable development (Vörösmarty *et al.*, 2010). One of the early definitions of water security was given by the Global Water Partnership (2000), which viewed it as every person having "access to enough safe water at affordable cost to lead a clean, healthy and productive life, while ensuring the environment is protected and enhanced". This ministerial-level meeting at the World Water Forum at The Hague stated that despite the huge diversity of circumstances around the globe, all nations desire a future that includes the goal of water security. The forum had a very broad understanding of this notion: "This means ensuring that freshwater, coastal and related ecosystems are protected and improved; that sustainable development and political stability are promoted, that every person has access to enough safe water at an affordable cost to lead a healthy and productive life and that the vulnerable are protected from the risks of water-related hazards." It then outlined a seven-step roadmap to achieving water security:

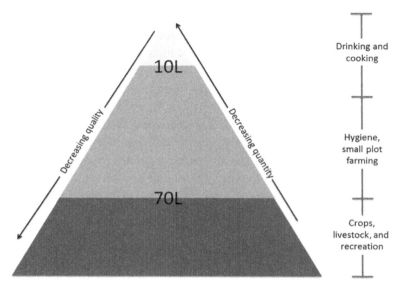

Figure 1.1 Hierarchy of water needs

1. Meeting basic human needs: this would be achieved through access, water quality and quantity, sanitation services which affect quality, and about participatory water management where the people, especially women, are empowered (Figure 1.1).
2. Securing the food supply: this would be achieved through more equitable allocation and efficient use of water, a process that is assumed to benefit the poor and vulnerable in society.
3. Protecting ecosystems: this would be achieved through sustainable water-resources management.
4. Sharing water resources: this would be achieved through cooperative "sustainable river-basin management or other appropriate approaches" at all possible geographic scales, the local, transboundary, or regional, by identifying synergies between different water uses
5. Managing risks: this would be done through protection from all water-related hazards ranging from pollution, to floods and droughts.
6. Valuing water: this would be achieved by gradually moving towards pricing water in a way that reflects the full cost of its provision, while respecting equitable access of the poor and vulnerable in order that they too can meet their basic human needs.
7. Governing water wisely: this would be achieved through the participation of all stakeholders in the management of water resources.

These understandings of water security are centered on people and ecosystems, with some being more explicitly eco-centric. The forum's definition, the most comprehensive of those reviewed, is so broad that makes its noble goals harder to achieve where, for example, people are expected to shed their human-centric view of water. In recent years, Grey and Sadoff (2007, p. 545) crafted the most quoted conceptualization of water security, which is, "the availability of an acceptable quantity and quality of water for health, livelihoods, ecosystems, and production, coupled with an acceptable level of water-related risks to people, environments, and economies". Grey *et al.* (2013, p. 4) later framed water security in terms of a "tolerable level of water-related risk to society". For Tindall and Campbell (2010, p.1), "water security is the protection of adequate water supplies for food, fiber, industrial, and residential needs". After explaining the notion, the same sentence explains that the goal "requires maximizing water-use efficiency, developing new supplies, and protecting water reserves in the event of scarcity due to natural, [manmade], or technological hazards" (see Figure 1.2).

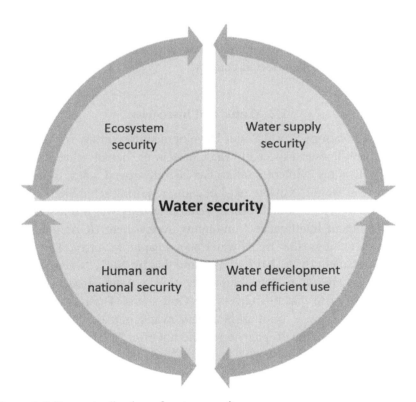

Figure 1.2 Conceptualization of water security

This book investigates the national security implications of the Gulf states' reliance on freshwater produced by desalination plants. This is done by assessing threats to their water and food security, and by suggesting ways to mitigate them. Since the early 1990s countries around the world have recognized that a multitude of national and transnational forces threaten their security and livelihood, and these threats include environmental degradation, disease, and climate change. Hence, the concept of security was expanded beyond traditional military threats against the nation state, and the notions of human and environmental security were developed (Jägerskog *et al.*, 2014; Swain, 2012). The Arab Human Development Report (UNDP, 2009, p. 2) defines human security as "the liberation of human beings from those intense, extensive, prolonged and comprehensive threats to which their lives and freedom are vulnerable".

These threats range from political instability in and around the Gulf states to environmental threats to desalination plants on the scale of the 2010 Deepwater Horizon oil spill which caused an environmental disaster in the Gulf of Mexico. Additional threats include major terrorist attacks on water or energy infrastructure disrupting supply for an extended period, or attacks that target skilled guest workers leading to mass emigration. While most of these are low portability scenarios, if any one of them were to occur, it would have catastrophic impacts on political stability of the affected country.

1.2 Water and insecurity

A recent United Nations World Water Development Report (WWAP 2012, p. 10) highlights a well-known aspect of water insecurity, namely that war interrupts water supply. It notes "violent conflict has also destroyed water infrastructure at different times in Beirut, Kuwait and Lebanon, requiring rehabilitation instead of expansion of delivery." Another form of water insecurity was discussed in 2012 by the US Government Intelligence Community Assessment (ICA), which focused on, among other things, the use of water as a weapon of terror. This thoroughly researched assessment (ICA, 2012, p. 3) found that by 2022, "water problems will contribute to instability in states important to US national security interests". Here, one can surmise that this covers the Gulf states. The report had a narrow scope focusing primarily on transboundary watercourses. The classified and declassified portions of the 26-page report did not offer "a comprehensive analysis of the entire global water landscape"; instead it "focused on a finite number of states that are strategically important to the United States and transboundary issues from a selected set of water basins", which are "sufficient to illustrate the intersections between water challenges and US national security" (ICA, 2012). Because the assessment's main focus was on transboundary watercourses, it paid cursory attention to aquifers and critical water infrastructure. For example, it found that until 2022, water in shared basins will be used by some states as leverage over neighboring ones. After that time, water shortages will become more severe, and water will increasingly be used to "further terrorist objectives (ICA, 2012, p. iii)".

1.3 Potential threats to water supply in the Gulf states?

Beyond 2022, violent extremists and rogue states will likely threaten to inflict substantial harm by targeting high-publicity physical infrastructure. The ICA assessment adds the following critical observation: "Even if an attack is less than fully successful, the fear of massive floods or loss of water resources would alarm the public and cause governments to take costly measures to protect the water infrastructure." It goes on to state that desalination plants and supply networks are likely targets for terrorists (ICA, 2012, p. 4). The assessment reduces factors that contribute to water-related destabilizations to:

(1) countries with inadequate water supplies, and which do not have the financial resources or technical ability to solve their internal water problems;
(2) some downstream states who have unresolved water-sharing issues and are "further stressed" by substantial reliance on water from upstream riparians;
(3) wealthier developing countries, which "probably will experience increasing water-related social disruptions but are capable of addressing water problems without risk of state failure"(ICA, 2012, p.3).

The assessment states that by 2022, "water problems – when combined with poverty, social tensions, environmental degradation, ineffectual leadership, and weak political institutions – contribute to social disruptions that can result in state failure" (ICA, 2012, p. iii). The report adds that the intelligence community has "moderate confidence" in their judgment partly because effective mitigation measures such as pricing mechanisms and infrastructure investments are available to these countries.

In discussing the MENA, the intelligence report argues that, "Increasing water shortages and rising food prices will present growing challenges for all but the wealthiest countries in these regions who can afford – typically with fossil fuel revenues – to subsidize food" (ICA, 2012, p. 8). The US Department of State requested the intelligence community's answer to this question: "How will water problems (shortages, poor water quality, or floods) impact US national security interests over the next 30 years?" The resulting assessment relied on previously published research and reports by intelligence agencies, and on "consultations with outside experts". The principal author of the report was the Defense Intelligence Agency (DIA), with contributions from other governmental agencies.

Table 1.4 *Net change in major land use around the world*

	1961 (Million ha)	2009 (Million ha)	Percentage net increase (1961–2009)
Rainfed	1,229	1,226	–0.2
Irrigated	139	301	117
Total (Cultivated land)	1,368	1,527	12

Source: FAO, 2011a.

While governments can always choose to increase water tariffs, such a measure is widely seen as being politically dangerous for any Gulf state to undertake. Further-more, the assessment appropriately acknowledges the inter-relatedness of social, environmental, governance, and water issues. However, the phrase "water-related social disruptions" implies social unrest that results from water issues such as chronic deficiency of freshwater supply that either makes it harder for the people to meet their basic needs or significantly erodes their quality of life. The report focuses on the government-to-people relationship and does not say much about the people-to-government side of the equation. Meaning, governments that fail in providing freshwater supplies to the masses would be viewed as having committed a significant breach of their social contract. Wealthier countries, the intelligence report finds, are well positioned to deal with water-supply challenges. Since desalin-ation technology was introduced to the Arab Gulf states, governments' emphases have exclusively followed supply-side management where they invested huge sums of money to establish new water and related infrastructures that are appropriate for a briskly modernizing economy and society. This development drive occurred just as these states were rapidly losing their mostly fossil groundwater reserves due to the introduction of modern, fuel-powered pumping technologies, largely unregulated water use, and mostly because of a heavy subsidy of the agricultural sector. Most fossil water in the Arabian Peninsula is 10,000 to 30,000 years old. Furthermore, aquifer water withdrawal in the Gulf states, mostly for irrigation purposes, is currently about six times faster than that of the natural renewal rate (Kumetat, 2012). This move by the Gulf states towards greater irrigation is somewhat similar to the global trend; the frequency of irrigation of cultivated lands around the world have more than doubled between 1961 and 2009 (Table 1.4; FAO, 2011a). In the Gulf states, most of the land that was irrigated had never been cultivated before.

1.4 Historical overview

The economic geography of production is such that countries tend to have a "home bias" during their early phases of development. When the Arab Gulf region came under British colonial rule in the early years of the nineteenth century, the population was small and primarily nomadic, and the economy was based on fishing, date-palm farming, and some trading. Until a few decades ago, the Arab Gulf region was the sleepy backwaters of the Arab world. The discovery of commercial quantities of oil in Bahrain by the Standard Oil Company of California (Socal) in 1932, and in Saudi Arabia in 1938, were the first, early indicators of the massive social and economic change that would soon descend on the people. The oil fields in the Arab Gulf were developed at different speeds. While Abu Dhabi began to export oil in 1962 and Oman in 1967, Kuwait developed its fields the fastest, which made it the largest oil producer in the region by 1953 (Metz, 1993). Kuwait was the earliest (1940s) to reap significant revenues from oil, which it used to vitalize the national economy, while Oman and Qatar (1980s) were late economic bloomers. In other words, by the early 1970s, the infrastructure of the Gulf states was minimal, embryonic, primitive, and wholly unsuitable for a modern economy. Therefore, as the oil wealth started to trickle in, leaders undertook the mammoth task of building an infrastructure for modern economies – almost from the ground up. Along with this came rapid improvements in people's quality of life, including rapid urbanization with all the services associated with that, and higher rates of water consumption; these necessitated building more desalination plants at larger scales. Kuwait's early endeavors towards economic development amount to a microcosm of other experiences in different countries in the region.

An essay with the title "Kuwait: a super-affluent society" (Shehab, 1964) might lead one to think that it is a commentary on current socioeconomic conditions; it is a five-decades-old essay, published in *Foreign Affairs*. It lists grand trade statistics, exorbitant expenditures on social services and on the country's economic infrastructure, and says that all this is "in sharp and dramatic contrast with the austerity of former times". Here, the timeline refers to the preceding two decades during which "the whole face of Kuwait has changed beyond recognition". The government spends more than $140 per person to provide freshwater and electricity, and "every tree and shrub that decorates her [Kuwait's] thoroughfares and public squares costs an average of some $250 a year". The government's expenditures on health, education, and other social services has "placed this tiny state on a higher level than some of the most sophisticated societies in the world; for in the fiscal year 1961–1962, it reached some $240 per inhabitant as compared with $210 in the UK and slightly less in Sweden".

The sudden explosion of hydrocarbon wealth upon the sleepy Gulf states has had profound and unprecedented impacts on every aspect of their society, economy, and environment. Even in the early 1960s, one key question that "has always been asked" by Kuwaitis is: how long is the spectacular wealth going to last? (Shehab, 1964). This same question continues to bear on the minds of all Arab Gulf states whose socio-economic condition was rapidly transformed from a state

Table 1.5 *Urban population growth (annual percentage)*

Country	2002	2003	2004	2005	2006	2007	2008	2009	2010
World	2.10	2.07	2.06	2.04	2.06	2.05	2.04	2.01	1.98
Bahrain	−0.07	0.79	3.73	7.60	11.33	13.23	12.87	10.61	7.64
Jordan	2.42	2.47	2.41	2.27	2.34	2.28	2.23	2.24	2.26
Kuwait	2.97	2.74	2.93	3.37	3.81	4.04	4.05	3.79	3.38
Oman	1.01	1.40	1.77	2.10	2.54	2.85	2.97	2.87	2.61
Qatar	2.72	4.70	9.12	13.91	17.62	18.67	17.05	13.58	9.69
Saudi Arabia	4.01	4.28	4.26	4.22	3.82	3.51	2.37	2.23	3.97
Turkey	2.18	2.14	2.12	2.12	2.02	2.01	1.99	1.96	1.92
UAE	3.26	4.37	7.26	10.63	13.69	14.86	13.90	11.23	8.01

Source: World Bank Data. URL: http://data.worldbank.org/.

of poverty, stagnation, and obscurity to being regional and global actors in a dynamic and complex world economic order. They, like Kuwait, have been feeling the need to build social and physical infrastructure, develop an industrial base, and set up public and financial services – all in a short window of time.

1.5 Lateral pressure

Pressure on natural resources available within a country fuel its outward expansion. The hydrocarbon boom that the Gulf states have been experiencing gave rise to gigantic development ambitions, which necessitated leaders to initiate sweeping and comprehensive development and modernization projects. They required these states to import labor, technology, and know-how from different countries around the globe. This is germane because domestic water consumption is affected by factors such as population size, frequency of water supply, and people's consumption patterns. The latter is affected by the quality of life, per capita incomes, and urbanization in the Gulf states (Table 1.5).[1] Urban population grew at a much faster rate than the national population, reflecting the rapid rural-to-urban migration, and the high levels of international immigration to the Gulf states. The vast majority of the latter ended up in cities. The Gulf states decided to invest in farms in other countries where they produce food crops to be imported to the investing state. They deployed the tools of globalization to boost their food security by expanding their foodshed beyond their national borders. This has also allowed them to conserve what is left of their groundwater supplies, and to treat them as a strategic resource to be used in times of emergencies. Will this approach meet its desired objectives? The test will come when food prices skyrocket again, or when there is a period of

---

[1] Additional factors such as households' distances from the water source affect domestic consumption in poorer countries like Yemen and Sudan (Roudi-Fahimi *et al.* 2002).

political instability in the wider Gulf region or a major food-exporting country. Either way, food security is likely to remain a high-priority item on the political agenda of the Gulf states.

In his famous (1968) book, *Political Order in Changing Societies*, Samuel P. Huntington argued that modernization and economic development generate intense social change, and that "societies in the throes of dramatic social transformation ... tend to be unstable and even violent. Positive outcomes are likely to emerge only where healthy political institutions capable of channeling and responding to such changes exist – and building such institutions is an extremely difficult and time-consuming task..." (Berman, 2009).

If and when countries develop such functional institutions, they can act as threat minimizers (Renaud and Wirkus, 2012). Gulf economies have long outgrown their human and natural resources. Different forces converged to create demand-induced scarcities, which pressured them to expand their resource capture beyond their national borders. For example, for decades before the advent of desalination plants, Kuwaitis imported water from southern Iraq. Conceptually, the lateral-pressure theory foresees a clash between countries experiencing resource scarcities. It argues that individuals and societies have a tendency to extend their reach and influence beyond known boundaries, whether they are economic, political, scientific, religious, or otherwise (Choucri and North, 1975; Gleditsch, 2001). Lateral pressure is similar to what had been referred to as "economic expansion" (Sorokin, 1957) or "outward expansion" of countries (Kuznets, 1966). When domestic capabilities are (or become) insufficient to meet demand, they generate lateral pressure where people, governments, or firms seek new capabilities beyond the national borders. Territories in which a state develops a "stake" could be perceived as falling within its sphere of national security and, therefore, worth defending. This outward expansion to capture or influence markets and natural resources, especially if it is combined with unequal access to resources (structural scarcity), increases the likelihood of hostile interactions with other competitors if actions are perceived as being dangerously competitive, threatening, or coercive (Choucri and North, 1975; Gleditsch, 2001). The expanding states that are most conflict-prone are those with high populations, high levels of technology, and inadequate resources (North, 1984; Choucri and North, 1989).

1.6 Cultural norms and water management

The familial monarchy political systems in the Arab Gulf have similar social contracts, where the central government shares revenues of oil largesse, and people enjoy tax-free incomes and cradle-to-grave social welfare systems in return for political acquiescence (Davidson, 2012; Barrett, 2011). For example, by 1994, the

Table 1.6 *General considerations for valuing water by types of usages*

Category of water use	Water's valuation
Water as biological need	Provided regardless of ability to pay
Water as spiritual need	Provided at a modest charge
Water as commodity	Provide at a full-cost recovery

Abstracted from Priscoli (2012); Ward and Michelsen (2002).

Kuwaiti government was subsidizing water and power to the tune of $1 billion a year (Cordesman, 1997, p. 58). Consequently, such a generous welfare system has nurtured a cultural mindset, in just a few generations, that views free or subsidized goods and services as the citizens' share of the hydrocarbon wealth, with no incentives to conserve. Why should an environmentally conscious citizen reduce water consumption if others in society will continue their wasteful usage? To be sure, assigning a single economic value to water is a tough challenge because desperate people will pay a lot to acquire a small amount of water.[2] Therefore, developing a framework for assigning value to freshwater "requires a clear statement of what the policy decision aims to achieve. The economic value of water measures the contribution of that water to accomplishing that decision's aim" (Ward and Michelsen, 2002, p. 443). Broadly speaking, the economic value of water is affected by its quantity as well as quality, and by its location and the time when it is in demand. One should also consider the different perspectives on the value of water, and account "for the difference between total, average, and incremental values of water" (Ward and Michelsen, 2002, p. 423) (Table 1.6).

If unchecked, the current wasteful behavior in the Gulf states could lead to some version of the tragedy of the commons. Furthermore, some aspects of water issues in the Arab world continue to be misunderstood even at the highest levels of government and academic circles in Western countries. Islam, the region's dominant religion, is often misconstrued. For example, a recent best-selling book on water (Solomon, 2010, p. 378) contends that because Islam views water as a free resource, "many Muslim countries charged little or nothing except partial delivery costs in some of the driest parts of the world." This position assumes that Muslim-majority countries apply Islamic law which they do not. Of the 1.7 billion Muslims worldwide, only Saudi Arabia and Iran are theocracies. Their combined population size is around 115 million. On the same point, a US intelligence report claims that, "In some portions of the Middle East, generation of financial revenue to make investments for basic water needs is limited by moral beliefs that water cannot be sold and only treatment and distribution costs may be recouped" (ICA, 2012, p. 10).

---

[2] Or to rid themselves of floodwaters.

While revenue generation is in fact "limited" in some countries, this is unrelated to "moral beliefs," a phrase that is used here as a code for Islam. Islamic jurisprudence offers detailed, nuanced water management guidelines for many different circumstances. For example, a person may obtain surface and subsurface water from the source at no charge if there are no associated costs (i.e. this does not apply to desalination) and as long as this does not impede the access of others in the community. Furthermore, all people have the right to quench their thirst, and therefore Muslims are required to share excess water to save fellow human beings. After people, priority should be given to meeting the needs of animals and then to plants. This ordering offers a roadmap on how to allocate water resources in times of water stress (Farouqi, 2001). Faruqi (2003, p. 210; see also Sadr, 2001; Wickström, 2010) argues that Islam views water as a community-owned social good and a fundamental human right, and views the natural environment as having considerable rights to water. He also concludes that Islamic law (sharia) supports water tariffs for the purpose of cost recovery, and supports involvement of the private sector in "service delivery, and up to full-cost recovery for water and wastewater services" as long as it falls short of private, exclusive "ownership over significant public water resources, or even long-term water use right". Jonathon David Walz (2010) concludes that all four schools of Islamic jurisprudence allow the sale of water supplies; the Maliki and Shafi'i schools allow for sale of unlimited amounts, while the Hanifi and Hanbali schools allow the sale of limited volumes. In addition to the theological principles, Muslims practiced water trade as recently as in the twentieth century. From 1925 to 1950, Kuwaiti corporations used fleets of dhows[3] to import freshwater from Shatt al-Arab in Iraq, some 100 km northwest of Kuwait city (Woertz, 2013; Crystal, 1995). In short, Islam views water as both a commodity and a common good which society must protect and preserve for current and future users. Furthermore, Gulf governments heavily subsidize freshwater supply and consumption, but for reasons that are related to social contract and governance – not to Islamic law. These governments have been willing to forego short-term losses for the long-term benefits of building politically, economically, and socially stable nation-states.

Commenting on the social contract in the Arab Gulf states, Raymond Barrett (2011) describes the flow of oil wealth to society as "river to the people". However, if it fails to satiate the needs of most or dries up, then the people may revolt. For him, Bahrain's popular uprising of 2011 was not so much a sectarian or freedom-based revolt as it was a result of the ruling family's inability to uphold its side of the social contract.[4] The kingdom's small population base had benefited

---

[3] A dhow is traditional boat that people of the Gulf, Arabs and Persians, used to sail in the Gulf, Red Sea, and Indian Ocean. It has a low front and high back, and is typically powered by triangle-shaped sails.

[4] For a different, more nuanced perspective on sectarianism and the Arab Spring, see Al-Rasheed (2011).

from petrodollars. However, its net oil exports had dropped from 41,000 barrels per day in 1980 to 12,000 barrels in 2013. Also, since 1980, its proven reserves have been cut in half to reach 0.124 billion barrels in 2014 (EIA, n.d. b). Given Bahrain's political upheaval and waning economic prospects, the GCC governments moved to assist it (and Oman) with $20 billion in short- and medium-term financial aid (El-tablawy, 2011). In a similar preemptive move during the Arab Spring, the UAE's central government allocated $1.55 billion to help upgrade the electrical grid and water connections in the poorer, less developed emirates in the federation (El-tablawy, 2011). Subsequently, it added that it does not plan to raise its heavily subsidized rates for water and electricity, despite soaring consumption (Utilities ME, 2013a). The government of Kuwait gave $3,500 to every citizen and increased civil-servant salaries by 115 percent, and the government of Saudi Arabia invested over $130 billion in job creation, salary increases, and in building some 500,000 units of subsidized housing, as well as many other services (Vidino, 2013; Knowledge@Wharton, 2011). For its nationals, Saudi Arabia offered a loan forgiveness scheme and established its first-ever program to provide unemployment benefits. Finally, the government of Qatar announced a 60 percent increase in salaries and social benefits for state civilian employees, and 50 to 120 percent increases to military staff, depending on rank. It also decreed a 50 to 120 percent increase in pensions for retirees, whether civilian or military. These benefits will cost the treasury $8.24 billion (Reuters, 2011a).

This use of government largesse to appease and placate the population is somewhat similar to the wheat self-sufficiency program that Saudi Arabia had from the early 1980s until 2008, a period when politically connected investors set up wheat and alfalfa farms that were heavily subsidized by the government. The Embassy of Saudi Arabia in Washington, DC (n.d.) asserts that, by 1984, the kingdom "had become self-sufficient in wheat" after which it "began exporting wheat to some thirty countries, including China and the former Soviet Union". This, the Embassy's website claims, is one of the country's "agricultural achievements". In the psyche of most Saudis, food (i.e. wheat) self-sufficiency is a worthy goal and one that enhances their overall sense of security. This policy, however, had overlooked the environmental and hydrological impacts as dwindling groundwater supplies, a byproduct of this policy, have in fact undermined the country's water security. Neighboring Iran has been going through a similar experience. Bozorgmehr (2014) paints a detailed and tragic picture of Iran's water shortages and related problems, and argues that they are "largely of its own making". She places part of the blame on the country's meager and poorly distributed precipitation, but argues that the vast majority of the country's problems are related to its rapid population increase, industrial growth in arid areas, and general mismanagement. An Iranian government official is quoted as saying "mismanagement has been far more damaging than drought". The author argues that "the biggest problem is a system of generous subsidies that has encouraged wasteful use" of water where people, especially farmers, do not have incentives to conserve it.

A government report by the Abu Dhabi Environment Authority (ADEA) stated that the current domestic water consumption in the emirate "surpasses natural water supply by nearly 26 times" and, assuming a business-as-usual approach, its ground resources will be totally depleted in less than 40 years (Emirates247, 2012). Saudi Arabia's policy reversal of 2008 amounts to an admission of failure and mismanagement. Although renewable water resources are very low in Saudi Arabia, the country uses fossil water to sustain its agricultural sector and burns a lot of energy to produce freshwater from saline sources. These measures, argues a Deutsche Bank report (Heymann, 2010, p. 10), do not make environmental or economic sense, but continue to be practiced partly because of security considerations: "a country does not want to have to rely on food imports if at all possible. But the acute scarcity of water in Saudi Arabia has brought about a change of heart. Officials say that the only source of wheat up to 2016 will be imports."

Saudi Arabia decided to take advantage of the forces of globalization which facilitates international trade and long-range food transport by globalizing its (virtual[5]) water-resource capture. This reliance on the international trading system is a profoundly daring policy measure for a security-minded country, one that is known for cautious political maneuvering. In addition to domestic policy recalibrations of this sort, most Gulf countries have been purchasing or leasing (potential) farmlands abroad, a practice that its critics refer to as "land grabbing." Such practices are evidence of change in how human societies manage their water-dependent sectors (Falkenmark *et al.*, 2004; Hoekstra and Chapagain, 2008). These investments are theoretically consistent because when a country loses its comparative advantage in producing a certain good, it would be better off economically if it embraces a policy reversal to become a net importer of that same good (Hubbard and O'Brien, 2012). A few decades ago, Rothschild (1976, p. 302) predicted the outward push (i.e. lateral pressure) that food security would produce. He wrote that in the longer term, wealthier Organization of the Petroleum Exporting Countries (OPEC) countries are likely to place a higher priority on food self-sufficiency, expand their own agricultural sector, or "to develop new sources of supply". In the mid 1970s, he also stated that the oil-exporting state of Iraq will import Egyptian peasant farmers and have them farm in the Tigris and the Euphrates watershed, and that "the Arab Fund for Social and Economic Development proposes to invest at least a billion dollars to make the Sudan ... into an agricultural hinterland for the Middle East." In other words, farming abroad is an

---

[5] Virtual water is water that is embedded in the production of food and other goods that people use.

old idea, and its original scope was water-rich countries in the Arab region – not beyond it.

As a response to lateral pressure, some Gulf states have been making techno-logical adaptations such as installing efficient faucets and irrigation systems. Furthermore, since the early 2000s, they have started exploring the possible future roles that nuclear technology may be able to play in their economic development. The UAE's Foreign Minister, Shaikh Abdullah bin Zayed Al Nahyan, said that his country's peaceful nuclear energy program is intended to reduce the growing demand for energy, which is rising at 9 percent annually. It would also help the country produce freshwater from nuclear-powered desalination plants (Khaleej-Times, 2011).

The lateral pressure to which the Gulf states responded allowed them to expand their resource capture to ensure their food and water security[6] – all the while, national policies allowed for a significant increase of residents' water and energy consumption thereby creating an unintended, boomerang effect of new sources of insecurity. In addition, there are other local and regional security risks that may destabilize the Gulf states, in turn threatening water-supply infrastructures.

1.7 A Gulf Spring?

The Arab Spring refers to the grass-roots peaceful uprisings by the people against their authoritarian leaders that were started in Tunisia in December 2010, and then spread into Egypt, Yemen, Syria, and other Arab countries. There were, for example, smaller, short-lived protests in countries such as Jordan, Oman, and Saudi Arabia. The people of Bahrain also rose up but were quelled by the security services who received military support from fellow GCC states. Until today, Bahrain experiences low-intensity agitation, which turns violent at times. While these region-wide uprisings turned violent in Yemen and descended into a pro-tracted and very bloody civil war in Syria, they produced a success story in Tunisia where the people drafted a new constitution and democratically elected a new government. They also generated political jitters among the ruling families of the Gulf states.

The massive deployment of financial resources starting in 2011, the first year of the Arab Spring, and the use of "an extensive patronage network" helped in containing social pressure, hence stabilizing the Gulf monarchies, and placating the people (Vidino, 2013; Kamrava, 2012); at least for the time being. In addition to this, the subsidization of utilities has created excessive use of water resources

---

[6] By importing desalination technologies from around the world.

and by extension, of energy resources. The dilemma is, however, that a recent Chatham House report concluded that "the systemic waste of natural resources in the Gulf is eroding economic resilience to shocks and increasing security risks" (Lahn, et al., 2013, p. vi).

The effectiveness of the material and symbolic resources that sustain the Gulf states may be short-lived; they may be eroded by market forces or by technological change. A sustained drop in oil prices would weaken governments' abilities to co-opt reform-minded nationals, especially the youth. Furthermore, people increasingly feel empowered by the social media because it allows them to circumvent censorship and access unfiltered news. The youth in particular are emboldened by independent, reform-minded imams who are sympathetic to their causes.

Al-Faisal (2011) argued that the turmoil in Arab countries, the regional aggressiveness of Iran during President Ahmadinejad's reign, the slow burning Israeli–Palestinian conflict, and "the creation and exploitation of terrorist enclaves" have had a profound influence on security and stability in the Gulf, which is inextricably linked to the issue of global energy. He also stated that Saudi Arabia is experiencing a "rising tide of nationalist sentiment that is binding the country together ever more firmly", so much so that a call for widespread protests against the monarchy, the "Day of Rage," fizzled, and effectively did not materialize. The former intelligence chief, however, did not mention the strict measures that were taken to preempt the protests: security forces who were deployed in huge numbers, blocked roads leading to the designated public square where people were to gather. The hundreds who dared to show up were harassed, fired upon with rubber bullets, and many were arrested. The protesters called for "increased democracy" in their country (Birnbaum, 2011). Also, the "rising" sense of nationalism is a backhanded acknowledgement of the weak sense of national identity where, for some, allegiance to a prominent Muslim theologian sometimes takes precedence over those of national leaders. This is part of the unresolved tension in the Middle East and North Africa between state-centric (nationalism) identities and those that are transnational (Islamism).

The Gulf states have been stable mostly because their native population tends to be small hence easier to manage (or co-opt), the rentier system allows for lucrative government employment and massive patronage opportunities, and the monarchical systems cast a shadow legitimacy on the ruling families. This situation has been referred to as "Gulf exceptionalism" because it appears to dissuade the people of the Gulf from embracing the popular uprisings. However, some leading Gulf officials and members of ruling families cast doubt on this assumed exceptionalism. Kuwait's prime minister, Sheikh Nasser al-Mohammad al-Sabah, warned that his government would have "zero tolerance" for anyone threatening Kuwait's security. A few weeks later he acknowledged that reform in all the Gulf states is a crucial step because "It is not possible to realize growth and stability in any country without economic and political reforms and to realize welfare of the peoples." (Reuters, 2011c). Similarly, the prime minister and foreign minister of Qatar, Hamad bin Jasim bin Jabr Al Thani, said in 2011 that relations between the people and their rulers are stronger in the Gulf states than in other countries in the region, and yet he observed that "I can't say that GCC countries are hundred per cent immune" to the currents of the Arab Spring (The Peninsula, 2011b).

Marc Lynch (2011) observed that during the Arab Spring, the GCC was able to drive the regional agenda because its surplus of petrodollars buys its leverage. He, however, argues that the organization's "power rests on much shakier foundations than is generally recognized" because "its internal divisions will likely re-emerge, its domestic political stability likely won't last" the ongoing turmoil in Yemen, and Bahrain's bloody suppression of its domestic opposition, its "sectarianism, and ongoing repression will continue to poison the Gulf from within". Lynch doubts the GCC countries' ability to maintain their regional leadership. Given that the eruption of the Arab Spring surprised almost everyone, Gause (2011) argues that regional analysts overestimated "the stability of Arab authoritarianism". For many in the Middle East, "political freedom outweighed economic opportunity". It was widely assumed that beneficiaries of authoritarian regimes would support them. The reality was very different, "the state-bred tycoons either fled or were unable to stop events and landed in post-revolutionary prison. The upper-middle class did not demonstrate in favor of Ben Ali or Mubarak. In fact, some members became revolutionary leaders themselves" (Gause, 2011).

The Arab Spring has had a profound and transformational effect on the people of the Middle East and North Africa. It shattered, for example, the long-held perception that heads of state and their security apparatus are too powerful to topple, and more citizens are calling for the right to participate in their own governance. For example, the political beliefs of Salman Al Awda, one of Saudi Arabia's most prominent imams, have evolved over the years and he is currently a populist promoter of democracy and civic tolerance. Al Qaeda's violent attacks in the kingdom between 2003 and 2005 along with the Arab Spring uprisings appear to have shaped his intellectual and theological ideas into what they are today. Although this is consistent with "the slow liberalization in Saudi society", the Saudi political system has largely remained static and unresponsive (Worth, 2014). Al Awda asserts that the "Gulf governments are fighting Arab democracy [in Egypt], because they fear it will come here" (Worth, 2014). They spent billions of dollars to subvert the popular, anti-regime uprising in Egypt and turn it into a "Gulf project" causing the "Saudi government to lose its friends and potentially lose its own people and invite disaster" (Worth, 2014).

In light of the Arab Spring, Al Awda wrote a book in which he invoked Islamic jurisprudence and history, as well as ideas of leading Western intellectuals such as Machiavelli and Rousseau, to argue that (1) the only appropriate and rightful form of government is democracy, (2) theocracy is un-Islamic, (3) the separation of religious and political powers is mandatory, and (4) "the worst form of despotism is that practiced in the name of religion". In the conservative Gulf states, the values that Al Awda's arguments uphold are certainly uncommon, some would consider them to be radical, and perhaps even revolutionary. While his book is banned in his country of birth, it is widely available on the internet. Furthermore, it is believed that the imam's popular MBC television program was banished because he supported the anti-regime protesters in Egypt and Tunisia that led to the successful removal of the heads of states there (Worth, 2014).

Religious scholars in Saudi Arabia led major anti-regime public protests that called for radical reform in the early to mid 1990s after the first Gulf War. They formed the Sahwa (or Awakening), the "Kingdom's largest Islamist movement, which blends the political ideology of the Muslim Brotherhood with local Wahhabi religious ideas" (Lacroix, 2011). Government forces arrested hundreds if not thousands of the protesters who were calling for radical reforms. This became known as the "Sahwa insurrection" (Lacroix, 2011).

The fact that charismatic imams like Al Awda are willing to take positions known to irk the government is not new in Saudi Arabia. In 2013, the kingdom cracked down on guest workers who had failed to renew their work permits. This triggered harsh measures by the authorities and multiple race-based confrontations between the foreign workers, natives, and security forces. Imam Al Awda criticized state's treatment of foreigners and called for greater tolerance, compassion, and for a more inclusive vision (Worth, 2014). This refreshingly new perspective is alien to many of the privileged Gulf natives who view low-skilled guest workers as socially inferior. The point here is that citizens, even ones from prominent social, religious, or political circles, are willing to speak their minds even if they contravene the government's official position. They, especially the youth, are technologically empowered and much better educated and informed about regional and global affairs than previous generations. This critical development is worth monitoring and investigating further because it could be an indicator of just how turbid and muddy contemporary socio-political currents are, a dynamism that has implications for security and stability for the Gulf states. Despite the upheavals and the peoples' pronouncements about their need for greater space for freedom, the social contract in the GCC countries was not fundamentally rewritten during the Arab Spring. This risky political gamble reflects unyielding political vision and rigid institutional frameworks that have served ruling families – thus far.

1.8 Conclusions

The Gulf states have deployed their hydrocarbon wealth to create countries that now boast modern infrastructures and social services that are mostly on a par with those in developed countries, and to ensure water and food security for the people. Their ambitious development plans have meant that the Gulf states have needed to import armies of skilled and unskilled guest workers. The perceived ill-treatment of the latter has led many of them to harbor grievances against their employers and host governments; this social pressure could erupt and escalate, or could be exploited by national and sub-national actors. This situation could become a potential threat to water security in the Gulf States. Chapter 2 discusses this as well as the potential threat to water security that socio-sectarian cleavages may engender, especially in countries like Bahrain, Saudi Arabia, and Kuwait.

A study by D'Odorico *et al.* (2010; see also Walker and Salt, 2006) finds that international trade amounts to an optimization of eco-hydrological resources worldwide; however, it also reduces the resilience of say, food-importing countries as it erodes the redundancy[7] of domestic systems. This is because it weakens the importing countries' ability to absorb shocks such as a stupendous crop failure in the producing country, political turmoil that may interrupt supplies, and domestic changes in available resources. International food trade also disconnects consuming societies from the natural resources upon which they, and producing societies, depend (D'Odorico *et al.*, 2010). The authors advocate an alternative to virtual water trade that is "based on the notion of water solidarity, whereby (1) long distance transport of food occurs mainly in times of crop failure and food shortage, and (2) it does not let the available resources exceed the carrying capacity that the region would have in periods with no drought". Furthermore, trade in virtual water is increasingly threatening water security of food-exporting countries (Vos *et al.*, 2014).

Gulf governments used their oil rents to also reduce the level of social dissent through coercion or co-optation, which appears to bolster their hold on political power. Energy endowments allowed these water-deficient countries to import additional food stuffs that allowed them to artificially increase their population base and their overall human carrying capacity (Hoekstra and Chapagain, 2008). Natural wealth also helped governments overcome water and food insecurity, primarily by building large desalination plants, subsidizing the farming sector at home, importing food products, and, more recently, by gaining access to arable land abroad. The effects of farming abroad on local populations, especially their food security, and on their natural environment have been sources of extensive

---

[7] Redundancy is when multiple components can perform the same function, which would compensate for a loss or failure in the system allowing it to act as an "insurance" of sorts.

public pressure on agricultural investors from countries like Saudi Arabia and the UAE. The international news media and some non-governmental organizations portray these investments as "land grabs," and often focus on the adverse socio-economic and agricultural impacts on local populations. Although land abroad was intended to be farmed and most of its yields shipped to investing countries, Chapter 3 shows that this plan is proving to be more complicated and less promising than was first anticipated. Sustainable security occurs when both actors become more – not less – secure as a result of their economic engagement.

Since around the mid 1990s, the Gulf states have been, to varying degrees, addressing water consumption in open and sometimes aggressive ways by taking some concrete steps to conserve water, a scarce natural resource that is expensively supplemented by desalination. Chapter 4 discusses such efforts and finds that the approach has focused on technological fixes and voluntary conservation messages; that is, they have largely steered clear of using conservation pricing as a mechanism of public policy. Finally, Chapter 5 outlines potential demand management opportunities and argues for the need to introduce gradual and incremental pricing steps that would eventually approach full water cost recovery. Furthermore, it argues that marginalized guest workers who speak neither English nor Arabic may well hinder responses to disasters at times of major, extensive emergencies. For Gulf societies to be stable and secure, governments need to develop creative laws that are inclusive of guest workers and give them civil and political rights. This will enhance the stability and cohesiveness of these societies, and bolster their abilities to respond to national emergencies such a major water-supply disruption that may take days or weeks to restore. Countries like Qatar, the UAE, and Kuwait, where guest workers make up over 50 percent of the population, would be major beneficiaries of more enlightened expatriate policies.

2

Threats to water security

2.1 Introduction

Historically, people have moved from place to place driven by economic, political, cultural, environmental, and other factors. This movement may be within the same locale, region, or across multiple regions, and it may be forced, self-initiated, or by invitation, affecting receiving and sending countries in various ways. In the modern era, with the demarcation of legally binding international borders, governments have introduced controls to regulate who they let in and out, which helps states maintain law and order.

In the past, it was often a challenge for central governments in the Middle East to control and govern peripheral areas where nomads were plentiful. Early in the twentieth century, many Saudis lived a nomadic lifestyle where they roamed, often seasonally, in search of pasture and water. In 1910, King Abdul Aziz started to assert his control over the vast country and to "establish peace throughout the Kingdom" of Saudi Arabia by initiating a program of switching nomadic Bedouin tribes to sedentary farmers who farm the land and reside in permanent homes. The first sedentarization center was established in 1912 in the Ar-ta-wiyah where more than 200 thousand people were provided with housing accommodations (Shenaifi, 2013, p. 118). Settled nomads were easier to control. It was also easier to instill national loyalty in them, and to manage any security challenges that they may pose to the nascent regimes.

International relations, war, and population movements have been inextricably linked for centuries, if not longer. In some instances, countries have actively invited professionals and unskilled workers to immigrate as a way to help the country realize ambitious development plans. For example, after World War II, Germany invited a large number of Turkish "guest workers," and after the oil boom of the 1970s, the Arab Gulf states recruited massive numbers of Arab and non-Arab workers. In recent memory in the Middle East, the creation of the state of

Israel turned hundreds of thousands of Palestinians into refugees. Similarly, violent conflicts – such as the Lebanese civil war of 1975–1990, the Gulf War of 1991, the Iraq War of 2003, and the Syrian anti-regime-cum-civil war of 2011 – all have killed large numbers of people, and turned millions of citizens into refugees.

The vast majority of the literature on population and violence focuses on national or transnational violence causing population movement. However, the literature pays scant attention to the circumstances under which some migrants become a potential national security threat to their adopted country as a result of grievances. For example, in the early years of the twenty-first century, the United States and some European countries experienced terrorist attacks that were perpetrated mostly by foreign nationals or, in some cases, by their offspring who were born or raised since childhood in their parents' adopted country. These incidents elevated the issue of immigration to the top of the national security agenda in the affected countries and beyond. Guest workers in the Gulf states have held persistent grievances about their work and living conditions, and have repeatedly demonstrated and sometimes rioted in protest. Natives view the overwhelming number of foreigners as a threat to national and cultural security. This chapter analyzes potential threats to water supply that the simmering unease of expatriates could cause, especially if a small group of aggrieved foreigners were to revenge their perceived oppression violently. Another threat under consideration here is related to the vulnerable dependence of Gulf states on water from desalination plants. If such complex technology breaks down or becomes the target of terrorist attacks, the effects could be catastrophic (Figure 2.1).

In an effort to protect their national identities from the perceived onslaught of the massive number of migrant workers on their culture, the Gulf states have chosen to deny them many political and legal rights. The living conditions and wages of these workers are sometimes inadequate. Disgruntled foreign workers, as well as politicized ethnic or religious groups, may want to act on the frustrations of perceived oppression by targeting critical infrastructures of the host country. The US Department of State's report on terrorism for 2011 shows that worldwide terrorist attacks numbered over 10,000 incidents impacting 70 countries, a five-year low as far as the total number. Despite this overall drop, the Middle East and south Asia suffered 7,721 terrorist attacks, or just over 75 percent of the total for that year. Over 66 percent of worldwide attacks struck infrastructure or facilities such as transportation systems and public places. The most recent data show a "sharp increase in the number of attacks directed at energy infrastructure, including fuel tankers, fuel pipelines and electrical networks" (US Department of State, 2012c). Concern over the vulnerability of the infrastructure increases when the delivery network is extensive, the dependency is high, and the security milieu is prone to domestic or regional turbulence. Also, because it takes a lot of energy to produce, treat, and distribute water, security of the critical infrastructure is

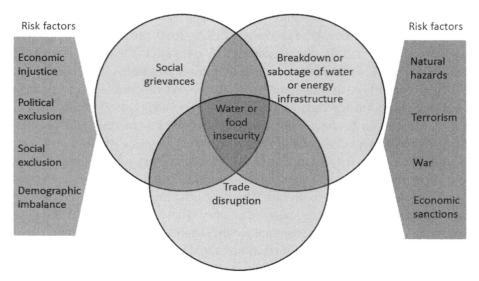

Figure 2.1 Threats to water and food security

interconnected. Ruhl *et al.* (2003) developed the following categories of the most serious types of attacks that a water sector is susceptible to: physical destruction of facilities, cyber attacks, biological agents added to a water supply, chemical agents added to a water supply, radioactive materials added to a water supply, and the release of hazardous treatment chemicals to the environment.

The current living conditions for many low-skilled foreign workers are below their expectations, a grievance that has contributed to recurring and animated protests. The Gulf countries have essentially two classes of foreign workers: those who hail from Western countries and those from developing countries. The vast majority of the former group tend to be professionals and are treated quite well, while the latter tend to be low-skilled laborers who are often treated poorly. Despite the comparatively low level of development that existed in Saudi Arabia six decades ago, the accommodations (and treatment) that it provided for Western workers then were superior to that of low-skilled workers today. The eastern Saudi city of Dhahran, once a "wildcat camp,"[1] was described in the 1940s as having "become a modern comfortable community. All the houses, offices, and community buildings are air-conditioned. The company provides its employees with water, electricity, gas, and telephones. Irrigation furnishes the basis for trees, grass, and flowers around the settlements. The employees have their own freshwater swimming pool, golf course, tennis courts, and club houses." The

---

[1] Wildcat camp is a term used to refer to a field where the drilling of a speculative oil well (or mine) is conducted and the productive potential of the area is unknown.

same amenities were also found in the nearby city Ras Tanura Dhahran, located on "a sandy coastal spit" (Stevens, 1949, p. 218–219).

The deficiency of water supplies in the Gulf states forced their governments to look for alternative sources. Desalination was their preferred choice and this significantly augments water supplies today, especially to major urban centers. Saudi Arabia's 36 desalination plants, located on the Persian Gulf and the Red Sea, are mostly powered by oil and natural gas. The country consumes 1.5 million barrels of oil per day to produce its freshwater needs (OOSKAnews, 2013). In the fall of 2014, its Saline Water Conversion Corporation (SWCC) said that the country's current production of about 2.6 million cubic meters of desalinated water per day will increase to 5.2 by 2016, and then again to 8.5 million cubic meters by 2025 (Saudi Gazette, 2014). Although the increase in production of desalinated water has been quite rapid, it is slower than what these countries have experienced in their earlier periods of their development. For example, Kuwait water production went from 3.1 million cubic meters per year in 1957 to 184 million cubic meters per year in 1987 (Murakami, 1995).

The inherent interdependence of energy and water production is perhaps stronger in the Gulf than in most other regions of the world because neither conventional water supply nor hydro-electricity are options, hence fossil fuels are burned to produce desalinated water. This relationship, sometimes described in terms of a nexus between water, food, and energy (Hoff, 2011), is being discussed at global summits of high-ranking political and economic leaders (Waughray, 2011) and at regional summits such as those of the GCC heads of state. As for intersectorial dependency, disrupting deliveries of drinking water or operations of wastewater treatment plants could have cascading effects on "public health, economic, psychological, and governance impacts on the nation". The core sectors with which water is interdependent "include agriculture and food, public health and health care, energy, chemical, dams, information technology, communications, transportation, and emergency services" and the critical manufacturing sector (Sullivan, 2011, p. 3).

Globally, the production of desalinated water stands at around 65.2 million cubic meters per day, with Saudi Arabia being the largest desalinating country (IRENA, 2012). The desalination process is energy intensive. The leading technologies are reverse osmosis (RO) and multi-stage flash (MSF) with the former accounting for 60 percent of the global capacity and the latter's share is 26.8 percent. The desalination of one cubic meter of seawater using a MSF process depletes 80.6 kWh of heat energy (290 MJ thermal energy per kg) as well as 2.5 to 3.5 kWh of electricity. Desalinating the same volume through a large-scale RO process requires an average of 3.5 to 5.0 kWh of electricity (IRENA, 2012; see also Siddiqi and Anadon, 2011). Given the substantial amount of energy required to produce freshwater, a disruption of energy supply would have a deleterious impact on the operation of desalination plants.

The Gulf countries share certain economic and demographic characteristics. In particular, they are rentier states whereby revenue from the oil sector is the dominant source of income. Fluctuations in oil prices have influenced the ebb and flow of foreign workers to the Gulf. An example of this is the drop in oil revenues in the mid to late 1980s which encouraged the Gulf corporations to replace Arab workers with less expensive ones from south and southeast Asian countries (Kassem, 1989). Another similarity is that the native population is Arabic by culture and ethnicity, and is predominantly comprised of Sunni Muslims, with the exception of Bahrain. The populations of these countries are small or very small (except for Saudi Arabia), and all are relying on extensive development projects to bring them into the modern age of globalization. Furthermore, their economies are heavily dependent on expatriate workers, the vast majority of whom are employed in the private sector.

According to Shah (2008), natives in five of the six Gulf Cooperation Council (GCC) countries considered the number of foreign workers in their countries to be too high and they would prefer to lower them. In Bahrain, the exception, people considered immigration levels to be satisfactory. While Bahrain's policy is to maintain the current immigration level, public opinion polls show that in all GCC countries, attitudes and policies are moving towards restricting the inflow of workers. The ratio of migrants to the total population of the Gulf states skyrocketed from 15 percent in the 1960s to 53 percent in 2010, which is many times higher than other large migrant-destination countries such as Australia, Canada, or the United Kingdom (Agunias, 2012). For decades now, the governments of the Gulf states have been struggling with how to manage the ever-growing number of foreign workers. For example, as the number of migrants has risen – sometimes rapidly – so has unemployment among locals. This phenomenon is sometimes called the "Gulf paradox," a term which refers to a country having a high domestic unemployment level while also relying heavily on foreign workers (Metwally, 2003 as cited in Chandra Das and Nilambari, 2009, p. 2). The strong growth of the Saudi economy has been generating jobs that are being filled by guest workers. hence the "high unemployment among the rapidly growing Saudi labor force". The overall unemployment rate is 5.8 percent and has remained broadly stable since the end of 2009. Between 2008 and 2014, the rate of unemployment among Saudis stood at almost 12 percent, or roughly double that of the overall rate. This is most pro-nounced among Saudi youth and women. An IMF report states that "Saudi employ-ment as a percent of total employment has declined since end-2009" (IMF, 2013, p. 14; see also ILO, 2011). The same is true for all the Gulf states whose net migration rate ranged from 2.1 percent per 1,000 people for Saudi Arabia, the lowest of the six countries, to Qatar and Oman, the highest, whose rates were 49 and 59 percent, respectively (UNDP, 2014).

Furthermore, Gulf governments have been under pressure from human rights organizations and lobbyists to enforce international standards with respect to workers' rights. These governments, on the other hand, are contending with unemployment among nationals, labor nationalization, and economic competitiveness. They are also faced with the transnational effects of the Arab Spring, which transformed the issue of unemployment of natives into a pressing national security concern, thus underscoring the critical role that job creation plays in political stability. The Gulf governments managed to resist or delay the Arab Spring contagion partly because their economies were able to bounce back relatively quickly from the global financial crisis which had pushed their economic growth to a low in 2009. After this drop, the gross domestic product (GDP) of the Gulf economies started to rise gradually and averaged 7.5 percent in 2011 (IMF, 2012a). The only exception to this was Bahrain, where political instability softened economic growth. Despite this, an IMF study argues that due to the rapid increase in the number of locals who are at the working age and the private sector's heavy reliance on foreign workers, "economic growth alone will not be sufficient to provide the needed number of jobs" for locals (IMF, 2012a, p. 20). Besides the pressure to create jobs, another area of concern for the Gulf states is that when complex technologies like those in desalination plants break down, the effects of their failure may be catastrophic, thus contributing to political and economic instability.

2.2 Desalination is defensible

Relations of asymmetric power between the GCC states and their neighboring countries in the wider region, coupled with tensions between them and Iran, have led Gulf Arabs to adopt desalination technology, which gives them the hydro-strategic advantage of an upstream riparian. While there have not been any major reported failures of desalination plants, there have been numerous incidents where critical infrastructure was targeted by criminals or terrorists.

Desalination technology places adopting countries in a better position than even an upstream state in a transboundary river system because it offers them all of its hydro-political advantages and more. A desalination-dependent state has the sovereign, exclusive, and complete physical control over its water source. In the case of Gulf states, this is critically important because they are heavily dependent on desalinated water. In other words, desalination for countries like Qatar or Kuwait is akin to France or the United Kingdom being dependent and sole owners of the

Seine or Thames rivers, respectively. In other words, reliance on water from desalination plants offers far greater geopolitical peace of mind than being a downstream state on a transboundary watercourse such as Egypt on the Nile River or Iraq on the Tigris and Euphrates rivers. The former is the last riparian on a river shared by ten others, while Iraq and Syria are downstream and mid-stream states, respectively, on the Tigris and Euphrates. The water shares of all three countries are, therefore, susceptible to the political whims and developmental vagaries of upstream states, not to mention natural pressures like droughts. Furthermore, a significant percentage of the labor force in a developing country is employed in the farming sector, which is by far the largest user of water. As a result, a reduction in the amount of water received would have an adverse impact not only on the economy, but also on the fragile social stability. Technologies that allow countries to produce water within national boundaries help them bypass what would be "upstream" pressure by another state, and therefore preserve their strategic independence and degrees of freedom in shaping their foreign policy.

This inward-looking, technology-centric security offers clear advantages but also raises questions about, for example, desalination's long-term financial and ecological sustainability, and about its vulnerability. Richard Perrow of Yale University writes that "even highly reliable systems are subject to everyday failures, and even if we avoid these, there is always the possibility of normal accidents – rare but inevitable in interactively complex, tightly coupled systems" (Perrow, 2011, p. 52). While the author was referring here to technological systems like those in Japan's Fukushima Daiichi nuclear power station that was damaged in a powerful tsunami in 2011, his comments should inform approaches to the security of huge and complex desalination systems as well. Perrow (2011, p. 51) adds, "It is much more common for systems with catastrophic potential to fail because of poor regulation, ignored warnings, production pressures, cost cutting, poor training, and so on." Human error is a known cause of accidents. Many desalination plants are designed in a modular fashion. Hence, a catastrophic system failure may be less likely, but the probability is not zero. It cannot be ignored.

Accidents and systems failures are all too common in modern technology-dependent urban centers. The normal accident theory (NAT) (Perrow, 1984; Perrow, 1999; Weick, 2004) states that under conditions of high interactive complexity and tight coupling, organizations may be disposed to accidents. This is likely to occur in systems (e.g. nuclear power plants, banking systems, or chemical plants) where two or more failures interact in unexpected ways, because components can have immediate and major impacts on each other. Safety training loses out to production pressures, human failings, and other reasons. The literature on NAT considers many different kinds of organizations and systems, but has not considered desalination plants as a key area of research. A study by Malik *et al.*

(1995) offers analysis on causes and remedies of failures in desalination plants in Saudi Arabia. It "describes the failure analysis carried out on the failed bolts used in the casings of an MSF plant seawater intake pump". A Kuwaiti study analyzed failure of desalination plants due to overheating (Husain and Habib, 2005). There is a real need for a greater level of research on the susceptibility of desalination plants in the climatic and geopolitical contexts of the Gulf states.

To minimize the adverse effects of water-supply interruptions, the security of the water-related infrastructure must become a central piece in emergency planning. This needs to be integrated at the community, provincial, and national levels because the water-related infrastructure includes electrical-power generation plants, pumps, and pipelines, among other facets. Also, there is a constant need to maintain, upgrade, and sometimes replace portions of the infrastructure. This requires investment at all levels from updating the skills of resident scientists and technicians to purchasing modern technology. During a recession or a period of low oil prices, budgets for water services in the Gulf states will likely be significantly reduced, which may introduce a vulnerability in the system for which government and societies would pay a heavy price in terms of economic decline and social instability. In this context, vulnerability refers to a nation states' ability to manage and adjust to exogenous stresses, be they sudden or gradual. This is influenced by institutional dynamism, good governance, economic strength, and diversity. In short, water emergency planning amounts to a "stress test" of how the entire system, from the elderly to first respondents, reacts in case potable water suddenly becomes unavailable for days or weeks.

Researchers in the areas of hydro-politics and climate change differ in their definition of terms such as "resilience," "adaptiveness," and "adaptation capacity," and they continue to debate these ideas (for an overview, see Gallopin, 2006). A resilient system, one that can bounce back after an incident, is an indicator of a high degree of water security in an area or country. A task force within the American Department of Homeland Security (DHS) describes resilience as the "ability to resist, absorb, recover from or successfully adapt to adversity or a change in conditions". In the context of critical infrastructure protection, the DHS defines resilience as the "ability of systems, infrastructures, government, business and citizenry to resist, absorb, recover from, or adapt to an adverse occurrence that may cause harm, destruction, or loss of national significance". It adds a precautionary, preemptive dimension to this conceptualization because resilience is also about the "capacity of an organization to recognize threats and hazards and make adjustments that will improve future protection efforts and risk reduction measures" (as quoted by Sullivan, 2011, p. 17). This comprehensive framework helps states reduce the vulnerability of their infrastructures from catastrophic collapse. To be sure, however, it is a tall order for young, developing countries like those in the Gulf. With the exception of Saudi Arabia, they were all colonies until the early 1970s, and modern national institutions were non-existent or very few. A state is considered resource vulnerable when certain issues make its political system susceptible to socioeconomic and political pressures or to natural shocks. On the other hand, a state is resource-secure when it is able to adapt and successfully respond to internal or external pressures or shocks. Normally, the development of a resilient system requires an advanced economy and well-managed, non-corrupt, and streamlined institutions.

Technological vulnerability and interdependence: illustrative case study

A recent illustration of water-resource vulnerability and interdependence occurred in Australia. A power outage caused an electrical surge and blew over 100 fuses, which irreparably damaged three motor-drive starters that forced authorities to halt operations of a water pumping station (Gilmore 2012a). The fuses had to be flown in from Sydney. The station distributes water that had been treated at the Bray Park Water Treatment plant to 75,000 residents in Tweed Heads, Banora Point, Tweed Coast, Murwillumbah and neighboring smaller areas. A local water and sewerage operations engineer (Gilmore, 2012b) said the "failure of fuses powering the pumps was 'unheard of'" and suspected that the initial problem had occurred in a distant location, "touched power lines and took out transformers". He also said that when the plant was built, it was one of the most advanced in Australia hence, "It was built for the future." Reflecting on the high-level surge protection placed on the facility, a local water manager said it was "just unlucky that all three (fuses) went at once". He then added that: "We had hoped the fuse replacement was all that was necessary to fix the pump station, however once the drives were energized, further significant damage was discovered."

Techno-optimists become heavily invested in the immense human benefits that technological advancements provide which makes them less sensitive to vulnerability. Learning about "normal" and induced accidents and failures will help water planners and decision makers introduce more informed policies and robust measures that are responsive to the needs of the community and the country. In addition to the potential for technological failure, extreme and unpredictable weather events are a threat to water-supply.

Tropical cyclones have historically been a common feature in the North Indian Ocean but have rarely affected the Arabian Peninsula. This pattern appears to be changing in terms of cyclones' frequency and intensity. Cyclone Gonu. the strongest recorded tropical storm to hit the Arabian Peninsula, was preceded by two similarly destructive storms which struck the region in 1890 and 1865 (Fritz *et al.* 2010). Since 2001, the three strongest tropical cyclones recorded have occurred in the Arabian Sea, including cyclone Gonu. Many of these countries like Oman and the United Arab Emirates (UAE) were not equipped to handle the effects of such storms. Flooding was especially devastating since many people live along the coast and in wadis that are susceptible to flash floods. Global warming is believed to be contributing to the frequency of these weather events in the Gulf states (WWAP, 2012; Kumar, 2009).

Most desalination plants in the Gulf states are located along the coasts of the Red Sea and of the Persian Gulf, the same region that is affected by cyclones, which potentially exposes the water infrastructure to tremendous risk. In the summer of 2007, cyclone Gonu made landfall along the coast of Oman, the UAE, and Iran. The storm developed in the eastern Arabian Sea and was upgraded to a tropical storm on June 2. By June 4, the storm reached tropical cyclone strength and category five status. The cyclone first hit the eastern tip of Oman at Ras al-Hadd, with winds of 164 km/h and the cyclone continued in a north–northwest pattern toward the Makran coast of Iran (Fritz *et al.*, 2010). Cyclone Gonu caused heavy rains, wind speeds up to 150 km/h, severe flooding, and rainfall reached 610 mm. Hardest hit were the Omani cities of Sur and Muscat. The storm caused flash floods, inundated the capital city with floodwater and mud, knocked down power lines and forced Seeb International Airport to suspend its flights. It killed 49 people, 14 remained missing, and over 20,000 were left homeless (Table 2.1). The cyclone damaged Barka and Gbubrah desalination plants, with the latter, Ghubrah, producing 42 million gallons of water a day and is the main source of freshwater for Muscat, a city of 631,000 people. The pipeline that supplies the Ghubrah power and desalination plant with natural gas was severed. Oman estimated the direct cost of this cyclone to be over \$4 billion (Al-Shaqsi, 2011). It took days to restore power to parts of the city, and after a week the Ghubrah plant was restored to 90 percent of its capacity (Mail and Guardian, 2007).

In 2010, cyclone Phet brought over 450 mm of rain to the northeastern parts of Oman. Damages were estimated at \$0.8 billion and 24 people died (Charabi, 2013). Phet was only classified as a category one storm but still did extensive damage to bridges, roads, electricity, water pipes, and desalination plants (Al-Shaibany, 2010).

The storm caused severe damage to power infrastructure that resulted in shortages of freshwater and electricity. Since a continuous supply of electricity is critical to the production of water, cyclone Gonu illustrates the interdependency between water and energy systems where power is critical to the production of freshwater. The World Economic Forum (WEF, 2011, p. 7) argues that shortages of water, food, or energy "could cause social and political instability, geopolitical conflict and irreparable environmental damage. Any strategy that focuses on one

Table 2.1 *Estimated impact of cyclone Gonu in Muscat*

Impact	Day 1 of cyclone	Day 3 of cyclone
Sheltered people	67,120 people in 139 shelters	2,650 people in four shelters
Affected roads	90 percent of all roads	20 percent of all roads
Electricity cuts	27 percent of the capital without power	4 percent of the capital without power
Water supply	23 percent of the capital without water	7 percent of the capital without water
Affected telephones	35 percent of network	1 percent of network
Affected mobiles	30 percent of network	3 percent of network
Fatalities	49 confirmed dead	14 still missing

Source: Al-Shaqsi, 2011.

part of the water–food–energy nexus without considering its interconnections risks serious unintended consequences."

Despite the effects of the high winds and the torrential rains and subsequent flooding, another serious concern of cyclones is the storm surge. Cyclone Gonu caused a storm surge that devastated the shallow coastline of Oman. There were high-water marks of 2 meters as far as 200 meters inland and these peaked at 5 meters at Ras al-Hadd. Surveys of the high-water marks along the coast illustrate the effect the storm surge had on Oman (Fritz *et al.*, 2010). While cyclone-related flooding can severely damage the electrical infrastructure and energy supply for desalination plants, it can also damage or destroy the plants themselves. Since desalination plants sit along the coast and the Omani coastline is particularly shallow, flooding of facilities can also prevent the basic operations of desalination plants. The storm surge could inundate freshwater storage tanks or the strong winds associated with cyclones could cause damage. During a cyclone, freshwater contained in storage tanks could be threatened by floodwaters that may contain sewage or hazardous chemicals. The high water levels recorded during cyclone Gonu could easily damage vital desalination infrastructure along the coast, the electronic equipment that runs pumping, and the intake pipes of desalination plants. Many desalination plants are not equipped to handle such flood levels. The water in a storm surge could also cause serious problems with electronics and circulation pumps due to saltwater. Extensive saltwater corrosion has been known to halt operations of desalination plants for weeks (Sheppard, 1994).

In addition to these technological challenges, many Gulf states are in the process of building nuclear plants for purposes of energy generation and desalination (discussed in a later chapter in this book). This spawns its own set of problems. A study by King Abdul Aziz University shows that radiation protection should be boosted in most establishments in Saudi Arabia where emergency plans and appropriate training on the use of technical instruments were absent (as quoted by Acton and Bowen, 2010). Acton and Bowen (2010, p. 454) find that the kingdom needs to "develop a comprehensive legal framework" and to "inculcate an appropriate safety culture", and to develop "national standards for radioactive waste disposal". The authors, however, were not optimistic about the future of the technology in the kingdom as they observed that "interest in radiation protection improvement was low". They also argue that while building certain components of the nuclear program is somewhat straightforward, it may be harder to develop the suitable safety culture and skills for managing radioactive waste.

An American expert on nuclear proliferation wrote that he was "deeply disturbed" by the "beginning of a Middle East nuclear arms race", sarcastically dismissing the need-of-energy argument that they have made to justify their pursuit of nuclear technology. "I have a hard time believing that Middle East leaders got together to watch Al Gore's movie and decided to reduce their carbon footprint. This is not about energy. It is about Iran." While countries in the region have the right to pursue civilian nuclear power plants, he warned that these are a "nuclear weapons starter kit", quoting Amory Lovins (Cirincione, 2009). The bulk of the highly skilled workforce are foreigners with temporary residency permits who have left their country of birth in pursuit of financial gain – not to adopt a new homeland that they would proudly serve with dedication.

2.3 Foreign workers outnumber natives

The GCC states had a resident population of 44.6 million people in 2010, of whom 21.1 million (47 percent) were non-nationals. While labor movement to the Gulf states has been ongoing since the discovery of oil, its pace was accelerated in the 1970s, and in some cases it occurred at a breakneck speed. Between 1992 and 2010, Saudi Arabia's population almost doubled as it rose from 16.9 million to 27 million. Much more dramatically, in just over three decades, the population of the UAE went from 0.55 million in 1975 to 8.2 million in 2010 (Shah, 2012).

The massive wave of migration to the Gulf states affects them in a number of different ways. First, foreign workers are perceived as a threat to national identity especially in countries like Qatar and the UAE which have experienced rapid population growth (Figure 2.2) yielding a yawning gap between foreigners and locals. Second, worker-receiving countries have had to expand their infrastructure for services such as education, health, water, and housing, as well as to provide amenities that people from diverse cultural backgrounds need in order to establish their new lives. Finally, the longer guest workers stay in a country, the more assertive they might become. This is facilitated by modern technology, mobile communication devices, social media, and by transnational rights organizations

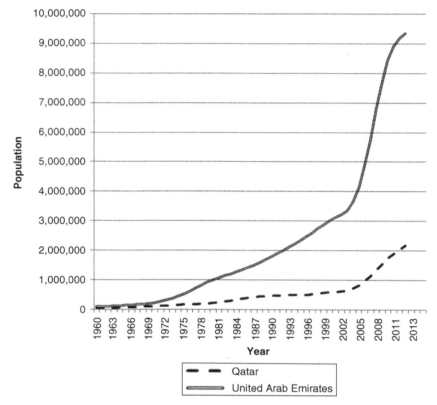

Figure 2.2 Population sizes of Qatar and the UAE, 1960–2013

who adopt the causes of oppressed groups. In short, Atiyyah (1996) argues that guest workers have "always been mistrusted by some Gulf nationals as a potentially destabilizing force and a source of unwanted influences on their political systems, societies and cultures" (Atiyyah, 1996).

Early signs of these concerns were evident many decades ago in Kuwait, one of the earliest Gulf states to develop its economy and infrastructure. Kuwait's labor experience may be considered a microcosm of the other Gulf states in terms of the range of issues and the policy responses to the demographic-cum-security challenges that they had to consider. Quoting Kuwait's first population census (1957), Shehab (1964) reports that the country's total indigenous labor force (15–60 years of age) was 23,977 of whom over 19,000 had either no training whatsoever or had minimal qualifications. Furthermore, some 4,000 were designated as "professionals and technologists," a group that included "only one chemist, one geologist, two physicians, two author-journalists, eight accountants and 156 clergymen" (Shehab, 1964, p. 465). The labor challenges that the Kuwaiti government had were obvious by the early 1960s; it had ambitious development plans, yet lacked the requisite human resources. However, as it invited larger numbers of skilled and unskilled foreign workers, it did not want to lose control over its own affairs (Shehab, 1964). The policy it decided to follow was to give employment "preference to its own citizens; and if suitable Kuwaitis were not available preference went to other Arab citizens and only in the last resort to non-Arabs" (Shehab, 1964, p. 465). Labor policies of the Gulf states were influenced by Kuwait's pioneering experience.

Guest workers supplied Kuwait with almost all its cadre of professionals from teachers, doctors, architects, and engineers, to administrative and managerial personnel, specialists who drafted its laws, built its industry, and established its financial institutions. The inflow of foreigners to Kuwait was so substantial that in the early 1960s, census data showed that natives were approaching minority status in their homeland. This labor rush affected, although not equally, all the GCC states; those with small populations were particularly impacted by the demographic shift. This "suddenly awakened" Kuwaiti natives and "alarmed" their decision makers who instituted "defensive measures" (Shehab, 1964, 468). The ratio of Kuwaitis stood at 47 percent of the total population in 1965 and stayed at this level until around 1974, partly due to "the mass naturalization" of Bedouins in the late 1960s (Russell and Al-Ramadhan, 1994, p. 569). The naturalization legislation of 1959 and its amendment in 1960 narrowed the definition of who was considered a native and hence deserving of Kuwaiti citizenship, and also allowed for naturalization of some foreigners, where Arabs were given preference over non-Arabs (Shehab, 1964, p. 468). It also set a quota of a maximum of 50 people per year to be granted citizenship, with exceptions allowed to Arabs.

In the 1980s, the demographic balance in the country was still a primary issue of concern and became the focus of Kuwait's Five Year Plan for 1985–1990, which aimed to have Kuwaitis comprise half of the total population by the year 2000. The plan also advocated increasing the population of Arab natives through natural increase (not naturalization), linked any naturalization to skills, and maintained that employment preference be given to Arab over non-Arab workers (Russell and Al-Ramadhan, 1994, p. 572–573). In the late 1980s, Kuwait's Supreme Planning Council drafted the country's Long-Term Strategy for Development, which included the goal of correcting the demographic imbalance. To do this, the council proposed the Selective Migration Policy (1989), which stated that migrants' length of residency should be "determined in consideration of national security, preserving Kuwait national identity, socioeconomic objectives, population composition, and government expenditures on services". It also argued that efforts to control immigration had been undermined by generous benefits and service packages that were offered to migrants, and by the fact that "real decisions about how much labor was needed and migrants' length of stay lay mainly in the hands of employers" and not the government (Russell and Al-Ramadhan, 1994, p. 573). With some variations, the experience of the other Gulf states with respect to the massive inflow of foreign labor has been similar to that of Kuwait.

2.4 Demographic imbalance and national security

About five decades ago, expatriates in Kuwait were referred to as being an "all-powerful community" because their influence permeated all aspects of life including locals' "cultural and artistic interest; their entertainments and dietetic habits; and even their colloquial Arabic" (Shehab, 1964, p. 467). This influence continues to be a concern for the Gulf governments in the twenty-first century. Their concern is centered on the large number in the foreign labor force, their concentration in certain strategic sectors of the economy, the effects that they are having on indigenous culture, and whether they will organize and act collectively in ways that may harm national security. Another consideration is that foreign governments or non-state actors may coax guest workers or nationals who belong to certain religious and ethnic communities to carry out acts of terrorism. Foreign workers were referred to in the Gulf and in academic literature as "aliens" and "transient immigrants" (Shehab, 1964), guest workers and expatriates, or by their ethnic and national background such as Indians, Egyptians, and Americans; this terminology is a continuous reminder to workers that they are not of the land or people where they reside.

Immigrants are also sometimes "described as a potential security risk" because they are seen as having divided loyalty towards their adopted country (Wimmer and Schiller, 2002). The Gulf states have varying concentrations of Shia nationals, many of whose ancestors came from Iran and are, therefore, ethnically Persian. The perceived threat that foreigners pose stems not only from their sheer numbers in comparison to the native population, but from their even larger percentage in the labor force, especially in the private sector. There are an estimated 10 million foreign workers in the Arab Gulf, of whom 70 percent are non-Arab Asians – a concentration that is as high as 87 per cent in the UAE (Agunias, 2012) and possibly higher than that in the private sector. This creates an extensive and substantial dependency that is almost impossible to reduce significantly without compromising the economic foundations of the host countries.

The overwhelming number of expatriates in the Gulf states and the globalized nature of Gulf states' economies are having tangible effects on native language and culture. At the United Arab Emirates University in the city of Al Ain, the number of students majoring in Arabic language and literature dropped from 600 in 2004 to 49 in 2013 (AlBahli, 2013). English is the lingua franca, if not the dominant language, among Arabs and non-Arabs in the Gulf states. While conversing with fellow Arabs online or in person, Arabic speakers routinely infuse English words into their vernacular, a common phenomenon that is captured by the term Arabizi (that is, the mix of Arabi and Englizi). Hence, preservation of national identity and reconnecting locals with their roots have become prominent items on the agendas of the governments of the region. The Gulf states' policies of limiting the rights of guest workers "stem from concerns about maintaining the national identity and national security, especially in countries with small populations and where citizens constitute a small minority of the population" (Ruhs, 2009, p. 21) such as Qatar and the UAE. Consequently, classifying expatriates as temporary migrants who have almost no opportunity of becoming citizens gives natives legal and numerical supremacy, which makes it easier for them to maintain their culture and national identity.

In addition to the demographic regulations mentioned above, governments over the years have introduced a sponsorship (Kafala) system, which limits the time that foreigners can stay, as well as a quota system for natives in different economic sectors. A migrant worker receives an entry visa and a residence permit only if a citizen sponsors him or her. The Kafeel, or sponsor-employer, applies for and receives an entry visa for a guest worker and becomes financially and legally responsible for him or her. During a foreign worker's stay in a particular Gulf country, he or she can only work for the sponsoring employer or institution, making that person completely dependent on the good will of employers in order to avoid deportation. This system also makes the employer responsible for the well-being and eventual departure of the employee. Granting ordinary people that much power over foreigners invites abuse wherein some employers impose extra charges on employees, and, in a few cases, do not pay them at all. Other immigration laws are intended to neutralize the cultural influence of foreigners by having them physically and socially segregated from natives. These measures have helped the countries to have tighter controls on the waves of foreign workers but have not slowed down their inflow; that is, the region's vexing labor dependency along with the frustrations of foreign workers and of unemployed locals continue unabated.

Human rights organizations have for some time focused a spotlight on the quandary of expatriates, which has caused Gulf states some degree of public embarrassment. Public pressure has, however, been intensifying because the economies of the Gulf states have become deeply integrated into the global economic system, which introduces a different set of expectations in terms of labor relations. For example, some Gulf countries have engaged world-renowned artists and architects to build iconic projects and buildings. Abu Dhabi, for example, is building a Louvre museum that is designed by Jean Nouvel, the world's largest Guggenheim museum (costing $800 million) by Frank Gehry, a maritime museum by Tadao Ando, Zayed National Museum by Norman Foster, and a concert hall that is designed by Zaha Hadid. By associating the names of famous architects and world-class institutions, the Gulf states become more vulnerable to international pressure regarding workers' rights.

Labor activists have been lobbying the British Museum and other western institutions who are involved in the project to press local authorities to improve workers' conditions. Qatar's story is similar. It was chosen as the host country for the World Cup in 2022. Expatriate workers' rights and conditions have been highlighted by human rights reports and by media outlets such as the British *Guardian* newspaper. Commenting on one of his organization's reports, the Secretary General of Amnesty International said that many migrant workers in Qatar "are being ruthlessly exploited" and said the "permissive environment and lax enforcement of labour protections" led to the large-scale exploitation of construction workers (Amnesty International, 2013). These complaints prompted Sepp Blatter, the FIFA president, to meet with Sheikh Tamim bin Hamad Al Thani, who emerged to say that labor laws will be rectified and greater attention will be given inspections of the workers' living quarters (Meenaghan, 2013). In 2013, prominent Saudi clerics, Salman Al Aoudi and Mohammed Al Areefi, separately called on their government to be more humane in dealing with expatriates, and to grant citizenship to those who are born in the kingdom (Emirates247, 2013). Combined, these religious personalities have around ten million followers on Twitter and major media presence, and hence they exert significant influence on public opinion. Similarly in the UAE, Sultan Sooud Al Qassemi, a member of the royal family in the Emirate of Sharjah, wrote in the *Gulf News* that authorities in his country need to find a path to grant citizenship to a select number of foreigners (Al Qassemi, 2013). His column generated an avalanche of comments and heated discussions on the newspaper's website. The Gulf states are clearly in the midst of a gigantic cultural change, and they are struggling to arrive at a demographic-immigration formula that is consistent with international norms, and yet meets their national goals and economic aspirations.

Public debate over workers' rights and living conditions, as well as techno-logical changes, are emboldening expatriates. In the summer of 2013, they rioted in protest of their poor working conditions, armed themselves with "metal spears and planks spiked with nails", and filmed their activism on a cell phone (Batty, 2013). As labor grievances persist, some event in the future is likely to act as a tipping point that will galvanize workers to escalate their protest by torching property or using home-made explosives to create physical damage and inflict psychological trauma on locals, other foreign workers, and on investors. According to one expert, Ekaterina Sokirianskaia of the Crisis Group, "modern terrorism doesn't require much resources – you need a man or a woman who are ready to sacrifice their lives, and explosives that you can actually make at home" (Flintoff, 2014).

The focus of this chapter now shifts to the likely implications if a few expatriate workers choose to act on their frustrations by targeting the critical infrastructure or, equally, if a subset of the native population commit terrorist acts motivated by a perceived injustice against their marginalized community.

2.5 Threats to critical infrastructure

The National Infrastructure Protection Center, which works with America's Federal Bureau of Investigation, defines a water-supply system as consisting of "[A] critical infrastructure characterized by the source of water, reservoirs and holding facilities, aqueducts and other transport systems, the filtration, cleaning and treatment systems, the pipeline, the cooling systems, and other delivery mechanisms that provide for domestic and industrial applications, including systems for dealing with water runoff, wastewater, and firefighting" (Ruhl *et al.*, 2003, p. 13). Because it is elaborate, extensive, and intricate, such a critical infrastructure is vulnerable not just to mechanical breakdowns, but also to human-induced failure or sabotage. The former is typically described as an accident, but not the latter – yet both disrupt drinking-water supply. While desalination plants may appear to be an autonomous modular arrangement, there are significant interdependencies linking political, economic, and technological systems together. For example, a significant amount of energy is needed to desalinate seawater or brackish groundwater; hence, an induced failure in an oil refinery or energy-delivery infrastructure would have devastating effects on water supply. Criminal elements and/or terrorists who plan attacks on critical infrastructure evaluate the attractiveness of a target using the following attributes (Veiga, 2011):

- criticality: public health and economic impacts of an attack;
- accessibility: ability to physically reach, enter, and leave the target;
- recuperability: ability of the system to recover from an attack;
- vulnerability: ease of accomplishing an attack;
- effect: amount of direct loss from an attack as measured by loss in production; and
- recognizability: ease of identifying a target.

The physical destruction of key components of the water system, which can be achieved by weapons, conventional explosives, and arson, is a more serious threat than chemical or biological attacks, primarily because obtaining the material for the latter is hard, and handling and delivering the hazardous material is a complex task. Attacks on critical infrastructures have been going on for as long as human societies have been around. It used to take localized, primitive forms, such a throwing dead animals in the enemy's water wells. In recent decades, however, these attacks have grown in their scale, methods, and in lethality. Iraqi forces which occupied Kuwait in 1990–1991 deliberately released three to four million barrels of oil into the waters of the Gulf, set over 700 of Kuwait's 800 oil wells on fire, and caused other wells to flow unchecked. Daehler and Majumdar (1992) argue that a motive for causing the sludgy oil slick was Iraq's interest in "depriving civilians and soldiers there (mostly in Saudi Arabia) of drinking water and also possibly of generation of electricity". The lasting damage was extensive, resulting in widespread ecological devastation. Given that the Gulf is a shallow body of water, the massive size of the slick and the threat it posed to the desalination plant supplying Riyadh, Saudi authorities shut down many of the plants as a precautionary measure.

Like other GCC countries, Kuwait's power-generation and water-desalination systems are fueled by, and therefore dependent on, oil and gas supplies. What saved the people was their country's post-war ability to produce "about 200,000 barrels per day, or just enough to keep people alive and maintain essential services" in terms of power and freshwater supplies (Levins, 1995). With this experience in mind, a colonel in the Kuwaiti military writes that "Kuwait's ports will remain within the range of Iranian and Iraqi anti-ship missiles" and that his country "will have to draw its water from easily targetable desalination plants" (Al-Samdan, n.d., p. 18).

Similarly, Saudi Arabia faces potential threats to its water infrastructure. Comparable to the deadly but unsuccessful 2004 attack on Yanbu, a major petroleum shipping terminal in Saudi Arabia, there was an attempted but failed attack on the Abqaiq oil facility in the winter of 2006. Al-Rodhan (2006) states that Abqaiq's importance is due to the fact that "nearly two thirds of Saudi Arabia's crude oil is exported" through this port. It "mostly produces Arab Extra Light crude, which requires little refining compared to other heavier crudes", and is the location of "the most important processing facility in Saudi and the world" where "crude is stabilized by controlling the levels of dissolved gas, natural gas liquids (NGLs) and hydrogen sulfide". While these attacks raised doubts about energy security, no public attention was given to the possibility of copycat attacks on the desalination infrastructure in the kingdom or elsewhere in the GCC states. Targeting national infrastructure should not surprise observers of terrorist groups because, for example, Al Qaeda has been preparing its foot soldiers for exactly that type of battle. Ahmed Ressam, a member of Al Qaeda, stated that the 1998 training that he and others like him had received in the organization's camps in Afghanistan included how to blow up "airports, railroads, large corporations" and how to wage urban warfare (Economist, 2001). In addition to attacking the coastal infrastructure of vital resources, terrorists targeted a loaded 160,000-ton Japanese oil tanker, M Star, in the Strait of Hormuz in July 2010. The ship's owner, Mitsui O. S.

K. Lines, and a report for the US Congress, believe that the resulting explosion was a terrorist attack, and a faction linked to Al Qaeda (Abdullah Azzam Brigades) claimed responsibility (Katzman, 2011, p.19).[2] During the Iran–Iraq War. both antagonists targeted oil tankers in the four-year Tanker War (1984–1988), and Iran blocked Iraq's oil exports from the Shatt al-Arab terminal. These experiences demonstrate that protection of the coastal infrastructure of the Gulf States is an integral part of their water security.

The issue of terrorism and piracy in the Gulf is of growing concern to American and Arab governments in the area. United States warships, in collaboration with naval forces of Arab Gulf states, seek to obstruct the movement of terrorists, weapons of mass destruction (WMD)-related technology and narcotics in the Gulf and the adjacent Arabian Sea. These forces also seek to contain piracy in the Arabian Sea. For the Americans, this is driven by national and strategic interests. The Gulf states concern is for the extreme vulnerability of critical high-capacity, long-lead-time replacement infrastructures such as energy installations and desalination plants.

A year before the 9/11 attacks in the United States, Al Qaeda used a speed boat to strike the American Navy destroyer USS Cole while at the port of Aden, Yemen. This method of warfare can be adapted and used by national and sub-national actors. Iran has increased its abilities to wage a guerrilla war. Since the early 2000s, Iran has been getting ready to wage warfare suitable for "asymmetrical" battle. For this, it relies heavily on the irregular forces of the Revolutionary Guards Corps (IRGC), which include a naval branch capable of attacking Gulf shipping and other soft targets. Connell argues that "In 2010, Iran had the largest inventory of ballistic missiles in the Middle East." He adds that their "limited accuracy suggests they would not be useful in a conventional counter-force role. Instead, they are probably intended for strategic targets such as cities, oil production and export facilities, ports and water desalinization plants" (Connell, 2010). Wilner (2011, p. 7) explains that "The development of lightly armed, low-tech, cost-effective weapons systems such as armed speed boats and seaplanes are intended to counter US hard power in the waters surrounding Iran using swarm tactics, and could very well be used to strike at US economic interests in the Persian Gulf, such as unarmed oil tankers and other commercial shipping." Since 2007, Iran has been increasing, modernizing, and upgrading its high-speed crafts with core missile and torpedo capabilities; some are assumed to be radar-evading (Himes, 2011). Speed boats laden with explosives can be used on the high seas like car bombs in urban settings, and can be more

---

[2] The width of the Persian Gulf reaches 21 miles at its narrowest point at the Strait of Hormuz where the truly navigable portion is even narrower. The strait has one shipping lane in each direction, and the bi-directional lanes are separated by a two-mile buffer zone. Therefore, disabling one or more ships in the strait could halt maritime traffic, which would impact food supplies to the Gulf states.

accurate and effective against their target than many missile systems. Furthermore, Iran has territorial disputes with the UAE, and tension between the Sunni and the Shia has been on the rise.

It should be noted that Iran's own energy facilities are also vulnerable to attack, and that the United States and her Arab Gulf allies are capable of retaliatory strikes against any aggression by Iran. While this is a real strategic vulnerability for the Iranians, the focus of the chapter is on the Arab countries of the Gulf. It is widely understood that Iran would retaliate if it were attacked by American or Israeli forces. The most likely areas to be targeted would be Arab Gulf states, and perhaps Israel.

Iran and Al Qaeda-central have been active in Saudi Arabia and Yemen since the 1990s. Colonel Salem Al Jaberi (2007) of the United Arab Emirates Army argues that Al Qaeda is likely to activate its so-called "sleeper-cells" to avenge for Gulf states' support of American policies in the region. He argues that they will likely target strategic installations that go beyond the oil and gas infrastructure (pipelines, refineries, and loading facilities) to include other offshore facilities like desalination plants. Other sleeper cells for Al Qaeda or Iran could exploit frustrations and grievances to advance their own political agenda and to expand their appeal among those who share in their anti-Western ideological outlook.

Terrorist attacks do not have to be overly complicated or expensive to be destructive. Some minimal degree of planning and analysis is sufficient to carry out a ruinous attack. About a year after the 2003 invasion of Iraq, suicide attackers in boats laden with explosives carried out three deadly coordinated strikes that disrupted Iraq's offshore Persian Gulf loading platforms. One targeted Khor al-Amaya and the others, the Basra terminal. While attacks on oil pipelines are somewhat common in post-Saddam Iraq, the targeting of these loading terminals was the first attack against the maritime assets of the country. Stratfor (2004) asserted "the platforms are perhaps Iraq's most important economic assets" and concluded "the significance of these attacks cannot be overestimated". The Basra platform was the only export terminal for the oil sector in the country. In a related note, Stratfor added "sabotage kept the Kirkuk-Ceyhan export pipeline – which exports crude north and then west to Turkey's Mediterranean coast – offline for 10 months after the war ended."

Due to the large size of the offshore loading platforms, "[A] successful strike would have to be precisely placed to do permanent damage to the massive platforms." Repairing them is complicated, and a successful attack "could damage it beyond repair" or take months to complete (Stratfor, 2004). Terrorists could launch similar attacks on offshore facilities such as a power installation or a desalination plant located anywhere in the Gulf or along the Red Sea coast. The Iranian coastline is long and jagged; small boats can easily hide along it. Also, there are active and somewhat routine smuggling activities between Iran and the

UAE. The Emirates, especially Dubai, are the center of much intrigue where Israeli intelligence, entrepreneurial nuclear scientists, and other shady operators do business. This is the place used by A. Q. Khan, the father of the Pakistan's nuclear program, for smuggling nuclear technology to Iran and Libya. The coastal topography or the illicit activities could be exploited to target a critical facility on the Arab side of the Gulf. Inflicting significant damage on a critical infrastructure that supplies the population with a vital resource would have incalculable impacts on the country's national security. In a 2008 memo to Washington, the American Embassy in Riyadh asserted that while desalination and power facilities may "not meet specific USG [US Government] interests", the Saudi government sees them as "critical to its ability to sustain essential services to its population." It continued, we "may wish to propose" a vulnerability assessment "of the Jubail Desalinization Plant (second on the Saudi's list of critical infrastructure after Albaqiq – in the original memo), an argument which likely finds favor with the Saudi government. The Jubail Desalinization Plant provides Riyadh with over 90 percent of its drinking water." The memo then commented on the likely impacts of a catastrophic failure at the Jubail plant, saying (Wikileaks, 2008) that "Riyadh would have to evacuate within a week if the plant, its pipelines. or associated power infrastructure were seriously damaged or destroyed. The current structure of the Saudi government could not exist without the Jubail Desalinization Plant." Metropolitan Riyadh is a modern city with the infrastructure and trappings that come with that status. The quality of life of its 6.8 million (2011) inhabitants has changed immensely; their grandparents or great grandparents drank from hand-dug wells until the 1930s. The simple lifestyle of that inward-looking society was far more resilient than its current manifestation that is tightly connected to the global economy and heavily dependent on a large number of guest workers.

2.6 Nationalization of the labor force

The GCC states' gradual nationalization of the labor force has slowly evolved from the initial ambitious goal of localization of all the jobs into prioritized efforts of appointing natives in strategically critical positions, and preventing concentration of workers from the same country in one sector. This is intended to ensure the continued functioning of critical infrastructure in times of crisis, especially crises pertaining to water-supply systems and other vital economic activities.

In order to prevent further expansion of the multicultural communities that expatriate workers have created, the vast majority of low-skilled and unskilled expatriates now live in secluded neighborhoods. Also, Gulf governments mandate that their citizens assert their national identity by donning traditional dress, a law that is enforced in the public sector, such as government offices, state universities, and all related enterprises. At the national level, Mohammed bin Rashid Al Maktoum, vice president and prime minister of the UAE and Ruler of Dubai, announced that 2013 will be the year when a "range of initiatives and policies will be launched to deal with the Emiratization as a national priority at all levels". He also said that the "most important characteristic of the founding fathers and the most important lesson we learned from them is that the citizen is the priority, and we should give importance to building humans before building edifices", and stressed that his short-term priority was "creating jobs for UAE nationals" (Khaleej Times, 2012). Furthermore, the UAE government approved measures[3] that were intended to inform nationals about their country and its history, help them take pride in their national economy, and ultimately build a strong national identity among the Emirati nationals (Khaleej Times, 2012).

Three years prior to the prime minister's announcement, Khalid Al Khazraji, a former undersecretary at the Ministry of Labor said: "The definition of Emiratisation as the development of national human resources is a reality. But if we mean replacing expat workers with Emirati ones, this is impossible" because Emiratis represent a minuscule percentage of the UAE workforce (Table 2.2). He added that the "concept of Emiratisation is meaningless and unworkable in the UAE labour market." Instead, he proposed that the country should focus on creating jobs for working-age Emiratis (Bitar, 2009). The Emiratization plan as outlined by the prime minister focuses on creating jobs for citizens, improving their skill set, and deepening their attachment to the homeland.

Governments of the Gulf states operate on the premise that employed citizens will be content and loyal and, hence, unlikely to be affected by the Arab Spring protests, whereby the people demand political reform and more rights. As a precautionary measure, Gulf leaders have recently lavished their people with various forms of financial incentives ranging from massive investments in affordable housing projects to very generous pay increases, and food subsidies. All these measures combined to help the Gulf monarchs sail through the storms of the Arab Spring without a significant political challenge. Bahrain, again, is an exception where social discontent simmers.

These measures and the prime minister's pronouncements amount to a re-definition of the concept and objectives of labor nationalization in the UAE.

---

[3] One such measure aims at integrating the "history of the Founding Fathers in the national curricula at schools". Another measure introduces a course on Emirati studies which will be required for students graduating from public as well as private institutions of higher education. The new course shall include an overview of the country's history, social culture, social and human geography, and legal system. The government decided to have an "Emirati Humanitarian Work Day" to commemorate the humanitarian work of the late founding father of the UAE Shaikh Zayed bin Sultan Al Nahyan. It also approved resolutions to prominently display the national flag throughout the country, in federal and local departments, and decided on a "Proudly Made in the UAE" initiative designed to promote national pride in locally made products (Khaleej Times, 2012).

Table 2.2 *Foreign labor and foreign population (1975–2008) as a percentage of the total*

	Foreign population (percent)				Foreign labor (percent)
	1975	1985	1997	2008	2008
Saudi Arabia	25	23	31	27	50.6
Kuwait	52	60	66	68	83.2
Bahrain	21	35	39	51	76.7
Oman	17	22	28	31	74.6
Qatar	59	60	67	87	94.3
UAE	70	79	76	81	85
Total					66.9

Table from Baldwin-Edwards (2011). Labor immigration and labor markets in the GCC countries: national patterns and trends, *Kuwait Program on Development, Governance and Globalization in the Gulf States*, March, no. 15.

This was implicitly driven by a sober reading of the demographic and economic landscapes, which show that conventional labor nationalization is a mirage. Short of this, decision makers appear to think that bolstering Emirati nationalism and the human capital of locals addresses the social concerns of citizens, and defines the political and legal framework that the government is using to manage the demographic reality for the long term. This approach is palatable because it does not require a major restructuring of the economy, which is needed to effectively reduce the dependence on foreign labor, nor does it demand much from citizens, and it will not disturb the rhythm of rapid economic growth that the country has been enjoying.

Despite the purported concern about the erosion of cultural and linguistic roots, many Arab families in the Gulf send their children to schools where the language of instruction is not Arabic. Their children also consume foreign-language entertainment online and via satellite television. This indicates that natives have a realistic understanding of the societal implications of hosting so many guest workers whose presence is but a manifestation of ambitious economic development plans, and that this mostly non-Arabic-speaking population will continue to be the dominant social group for the foreseeable future.

2.7 Demographic imbalance

The Gulf countries' demographic imbalance appears unsustainable. The population size of natives in some states is at around ten percent of the total, and this minority status of locals imposes legal, cultural, and political burdens on them. Some locals remind foreigners that they are transient residents whose stay expires with their contract, and view their presence through a security lens. For example, members of the governmental Federal National Council (FNC) in the UAE have described the loss of national identity and demographic imbalance as a "threat to national security" (Habboush, 2010). The multiculuralism literature finds that in some Western countries citizens also feel that "too much diversity" dilutes national identity, erodes the cohesion of society, and alters core cultural values (Vertovec, 2004; Wessendorf, 2008). In both Western countries and the Gulf states, the native population – which is usually the dominant group – describes immigrants as being alien to the community, a perception that justifies their marginalization. Immigrants are also sometimes described "as a potential security risk" because they are seen as having divided loyalty towards their adopted country (Wimmer and Schiller, 2002).

Such terminology, as well the housing conditions of low-skilled workers, spawn socio-psychological isolation and a distinction between locals and expatriates, which are re-enforced through unflattering narratives and representations. The result is the promotion of one's own culture and homeland while belittling that of "others," a binary construction that feeds social or racial tensions between the groups. Atiyyah (1996) argues that the soft legal status of expatriates has led some employers to deprive them of basic statuary rights, which "engender resentment among expatriates" and may "foster feelings of insecurity".

A preemptive acculturation measure that is argued to be excessive is Saudi Arabia's official discouraging of non-nationals from attending soccer stadiums as fans because it could lead "towards a greater attachment to their host country" (Dorsey, 2013). The idea is to deepen natives' attachment to their homeland and to discourage all other residents from developing similar feelings.

Before the first Gulf War of 1990–1991 when Iraq invaded Kuwait, Gulf "governments had followed an appeasement foreign policy", where they were reluctant to evict large numbers of Arab politically active workers "for fear of antagonizing" the more populous and politically stronger Arab governments (Atiyyah, 1996). The war constituted a watershed event in the history of migration trends to the Gulf states. Iraq's violation of a fellow GCC country created a political earthquake in the region and raised questions about the stability of political systems in the Gulf. The scale of the event emboldened the Gulf states into decisions that hitherto seemed unfathomable: to be in open military alliance with mostly Western armies, to use Gulf troops against a fellow Arab and Muslim aggressor state, and to take decisive action with respect to a foreign labor force.

The long-welcomed Arab workers, some of whom were naturalized citizens, became distrusted partly because of the allegations that "some expatriates" collaborated with Iraqi forces in occupied Kuwait (Atiyyah, 1996). Non-Gulf Arab residents are presumed to take the cue from their heads of state, or their heads of state are assumed to reflect the sentiments of the people they represent. For example, when the heads of states in Sudan, Palestine, Jordan, and Yemen sided with Saddam Hussein over the invasion of Kuwait, large numbers of Palestinians, Jordanians, and Yemenis living in Kuwait and Saudi Arabia, and Egyptians in Iraq became persona non grata, or, "enemies of the State", according to Lori (2012, p. 13).

Irrespective of personal positions on the conflict, over three million legal Arab immigrants were forced to leave the Gulf states (Lori, 2012). It was estimated that 700,000 Palestinians were living in the Gulf states of whom some 400,000 were in Kuwait. The latter group "constituted by far the most cohesive, successful, and politically conscious single Palestinian community in exile". Their economic strength and future in Kuwait was "abruptly and cruelly devastated" as their loss of income and assets were estimated as being "at least $8 billion" (Abed, 1991).

For decades, Yemeni workers had sought employment in GCC countries and were granted special privileges by the Saudi government. From 1958 until late in 1989, the Saudi government allowed them to enter and leave the country without a visa. This hospitality ended precipitously when the President of Yemen, Ali Abdullah Saleh, did not condemn the invasion and occupation of Kuwait. He took a position against the use of force in the liberation of Kuwait, a stance that reflected the "overwhelming public sentiment" of Yemenis (ICG, 2003). This political decision led the GCC countries to expel an estimated one million Yemeni citizens in the span of a year (Dorsey, 2010; Forsythe, 2011; ICG, 2003). Similarly, but on the opposite side of the conflict, Egyptians who had been working in Iraq were victimized. About 70 percent of Iraq's 1.3 million foreign workers were from Egypt; some of them were harassed and most were forced to leave (Russell, 1992) because Hosni Mubarak, the president of Egypt, had sided with the Gulf states and the United States.

In addition to the alleged security motivation behind the *en masse* expulsion of so many fellow Arabs, the action was mostly retaliatory in nature. The returning migrants and their lost remittances dealt a strong economic blow to the economies of their countries of origin, especially to that of Yemen, which had been scrambling to manage the nascent political unification of the war-torn country. Furthermore, the Gulf states stopped their significant financial assistance to the government of Yemen, and Saudi Arabia repealed the special status it had previously extended to 700,000 Yemenis and their families (ICG, 2003).

During the decades when Yemen was divided into north and south, Saudi Arabia was anxious about a "potential threat" of a strong Yemen Arab Republic. Hence, Saudi Arabia supported south Yemen (the People's Democratic Republic of Yemen), paid northern tribal leaders to encourage their independence from their central government in Sanaa, and supported "attacks on oil pipelines by Yemeni tribesmen" (ICG, 2003). This illustrates how a foreign government can use its economic and political resources to work with local non-state actors in another country, and to influence their allegiance and loyalty, in order to undermine the infrastructure and political stability of the target country. It also raises the question of whether a foreign actor could tap the frustrations of some workers or members of minority groups in the Gulf states, and get them to attack their home governments' energy infrastructure, affecting supplies to desalination plants and hence impairing water security.

2.8 Foreign workers after the Arab Spring

The weakest link in the water infrastructure in the Gulf states are the tens of thousands of pipelines that deliver water to consumers. Their extensive length makes them vulnerable to technical failure or deliberate attacks. While such disruptions can be repaired in short order, the same cannot be said about the damage inflicted on a major desalination plant that supplies water to a large number of people. Although this is a lower-probability event, it would have a much higher impact (Figure 2.3). In the event that such an attack failed, it would instill much fear and suspicion among the people as it would shake their faith in the ability of the central government to protect their well-being. However, if a "spectacular" attack was successful in inflicting significant damage on utilities infrastructure, and if it was followed by additional, easier-to-conduct strikes on softer targets (e.g. kidnaping and brutally killing Western expatriates), the social and political consequences of such terrorism would be difficult to contain. The professional expatriate class would likely be the first to depart, and this would exacerbate a Gulf country's ability to continue operating some of its complex water and energy facilities.

Similar to Kuwait's experience in 1990–1991, in Libya a small minority of foreign workers who did not flee the Arab Spring revolt of 2011 sided with the Gaddafi regime. Before the popular uprising turned violent, this oil-rich country had more than 2.5 million foreign workers, about one million of which were from sub-Saharan African countries. As the uprising against the regime progressed and intensified, much of the professional army either joined the rebels or stayed neutral. In the dying days of the regime, the ruling family bolstered its forces by resorting to "mercenary and praetorian forces" (Goldstone, 2011, p. 459; Hauslohner, 2011) and, in desperation, hired "African migrant workers as mercenaries" (Nkrumah, 2011). While the mercenaries were highly paid ($500 per day), many spoke neither Arabic nor English, which made it difficult for locals to reason with them. During and shortly after the uprising, scores of dark-skinned workers from Chad and Sudan were brutally murdered because they were suspected of being mercenaries for the regime (Quist-Arcton, 2011).

Levels of risk

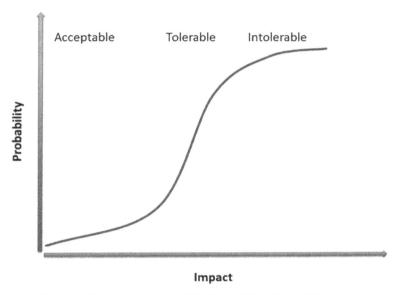

Figure 2.3 Relationships between probability and levels of risk

The sanctions that had been imposed on Libya by the United Nations and America had almost crippled the Libyan economy. When they were lifted in 2004, Western foreign direct investments started flowing and oil gushed out in much larger volumes, Libya was again able to modernize and re-integrate its economy into the global system. Due to this win–win arrangement, many of the Western and other voices that had called for and applauded the opening of Libya's economy remained silent on the issue of migrant workers' rights (Gurman, 2011).

The situation in the Gulf states has been somewhat similar. Short of human rights reports and the rare news or academic articles exposing how foreign workers are treated in the Gulf states, the governments of these countries have been counting on the West's passive response to continue. Furthermore, Ross (2001, p. 357) finds that in a rentier system, the general inclination of governments is to use "low tax rates and high spending to dampen pressures for democracy; a repression effect, by which governments build up their internal security forces." Conceptually, this helps in understanding why, in 2011, the UAE government signed a large contract with Erik Prince, the founder of the private military firm, Blackwater Worldwide, which he subsequently renamed to Xe Services. The company provided their customer, the Crown Prince of Abu Dhabi, with an 800-member battalion of foreign troops that would "conduct special operations missions inside and outside the country, defend oil pipelines and skyscrapers from terrorist attacks and put down internal revolts, the (leaked) documents show. Such troops could be deployed if the Emirates faced unrest in their crowded labor camps or were challenged by pro-democracy protests like those sweeping the Arab world this year" (Mazzetti and Hager, 2011). The foreign military force was initiated before the Arab Spring (Mazzetti and Hager, 2011), a fact that underscores the UAE government's apprehension about labor unrest and the security of its (oil) infrastructure. A successful attack on a major oil facility would interrupt fuel supplies and consequently affect production of potable water. It was also reported that Emirati officials are ready to pay billions of dollars to expand this force into an entire brigade of several thousand men but only after the current battalion proves itself in a "real world mission" (Mazzetti and Hager, 2011). This amounts to an official commentary on the low capability of the armed forces, which are made up mostly of foreigners. The armed forces of the other Gulf states are equally small and weak, with the possible exception of those in Saudi Arabia.

To summarize, given the absence of comprehensive measures to formally address labor grievances once and for all, Gulf states clearly believe that their regulations adequately respond to their needs at this time. This, however, underestimates the effects technology could have in helping workers mobilize and organize, and the fact that activism has become diffuse globally where public pressure and consumer boycotts have adverse effects on the business environment of the targeted country. The social media, together with technical support from cyber hacktivists abroad, was instrumental in the "success" of the Arab Spring as it proved to be an effective forum for helping ordinary citizens rebel against their rulers. Many of the unskilled guest workers are well aware of the grinding poverty in their country of origin and the prospect of unemployment if they were to return to it. Therefore, in case of sustained instability in their host country, a few would likely choose to work at high levels of risk by taking a side in the turmoil for purposes of revenge or for financial gain. Finally, the fall of the Mubarak regime in 2011 is a reminder that Western governments are willing to transfer their decades-old support of previous allies to the new political leadership if it is seen as serving their national and strategic interests.

The Arab Spring that started in late 2010 overthrew repressive regimes, and appeared to give Islamist movements the upper hand in the region. The Gulf states have watched these tectonic-level political changes with trepidation; some, like Saudi Arabia, argued that Western governments should have supported historical allies despite the mass popular movements against them. One such ally, President Hosni Mubarak of Egypt, was replaced by Mohammad Mursi of the Muslim Brotherhood. Since the party rose to power in 2012, it has consistently said that it has no plans to advocate political change beyond Egypt's borders. Mursi himself has said that his government has no plan to "export the revolution" (Reuters,

2012). These statements were primarily intended to calm the fears of the Gulf states. Despite this, serious tensions between the UAE and Egypt emerged. In 2012, the UAE arrested some 60 people for allegedly having ties to the banned Muslim Brotherhood and accused them of "conspiring to overthrow the government" (Reuters, 2012). Also, Saudi Arabia, the UAE, and Kuwait took a strong position in support of the coup d'état by the Egyptian military during the summer of 2013, which ousted Mohammed Mursi, the Muslim Brotherhood president.

In the wake of the Arab Spring, finding job opportunities for young nationals has become even more critical for social cohesion and political stability. Some six months after the start of the Arab Spring, Abu Dhabi's public sector and firms associated with it were ordered to find employment opportunities for natives. The popular uprisings that affected a few Arab countries were motivated, in part, by poverty and unemployment. This context helped push the addressing of unemployment among locals towards the top of the national security agenda; hence, governments activated labor nationalization programs (e.g. Emiratisation, Saudization, etc.). The literature on labor nationalization is rather limited (Randeree, 2012). The concept of nationalization is about "localization" where the government commonly sets policy that requires employers to gradually reduce or completely replace foreign workers by employing natives (Randeree, 2012; Zerovec and Bontenbal, 2011; Forstenlechner and Rutledge, 2010). All Gulf states have labor nationalization programs, with some in place for decades. Forstenlechner and Rutledge (2010) offer an apt summary of labor nationalization in the Gulf through illustrations from select experiences:

Bahrain sought to make nationals (relatively) less expensive for the private sector to employ by taxing expatriate labor. Oman was, and probably still is, the most dedicated to enforcing quotas and designating certain occupations to be staffed solely by nationals. Generally speaking, however, the results have been disappointing; in many instances, dependence on expatriate workers has continued to increase. For instance, during Saudi Arabia's five-year development plan spanning 1991–95, which set out to reduce the number of expatriates in the kingdom's workforce, they actually grew considerably.

An intensive effort targeting job creation has long been identified as a high priority for the Gulf states. About a decade before the Arab Spring started, an IMF report (Fasano, 2002) stated that creating job opportunities for the young, rapidly growing indigenous population of the UAE was critically important, and this could happen through education and training, as opposed to quotas. Economic growth reached a low in 2009 as a result of the global financial crisis, and the GDP of the GCC economies gradually rose to 7.5 percent in 2011. An IMF (2012b) report found that political instability during the Arab Spring lowered economic growth in Bahrain, and that it will take economic restructuring for the GCC economies to create the requisite number of jobs for their natives.

The private sector, the largest employer of foreign workers, has historically resisted labor nationalization laws because companies win work bids by delivering their product or service at the lowest possible price. This generally requires wage levels and employment conditions that do not appeal to natives. The private sector, for example, opposed the idea of setting a maximum stay limit (six years) for guest workers. Local business elites argued that it would increase their operational costs. However, the Arab Spring motivated Gulf governments to press employers to meet workforce nationalization targets by hiring more nationals. While it is not clear how effective this pressure was on the private sector, it was easier for central governments to create job opportunities for nationals. In the spring of 2011, the UAE government employed 6,000 jobless Emiratis (Kerr and Mishkin, 2011).

The number of locals who are employed in the private sector is negligible. Hence there is ample room here for expanding targeted job creation. However, this requires boosting the quality and modifying the focus of educational and training programs at the national or corporate scales, and giving targeted subsidies for employing qualified but inexperienced nationals. Closing the salary and benefits gaps between private and public sectors is critical in making the former more attractive to nationals who gravitate towards positions in public enterprises. Rebalancing the labor market requires a long-term strategy that seriously considers and gradually transitions towards structural solutions in which wages are related to labor productivity (Fasano and Iqbal, 2003).

Quietly, the UAE government significantly curtailed the number of work permits given to citizens from countries affected by the Arab Spring, fearing that they might be politically charged. The Gulf governments' decades-old fear of an "influx of low-skilled and politicized" workers (Ulrichsen, 2009) was amplified in the aftermath of the Arab Spring. In the early phase of their modernization, the Gulf states had placed a premium on racial and linguistic harmony in the homeland when they decided to recruit mostly Arab workers. These workers did much to help the newly independent states build modern institutions, infrastructure, and services, and they prospered themselves from opportunities that were often tied to relatively high salaries.

As a result of wars that occurred in different Middle Eastern countries, large scores of homeless Arabs migrated to the Gulf states (Russell, 1992). Given the circumstances of their departure, one would expect many refugees to have strong political views about who caused their plight. Over time, and especially after the invasion of Kuwait, the Gulf states became particularly sensitive to the politicization of Arab workers because they were seen as carriers of, or susceptible to, unwelcome political and social ideologies such as Communism, secularism, Arab nationalism, and Islamism. The latter two discount current political boundaries as being colonial artifacts and, accordingly, advocate a regional distribution of natural resources including wealth generated from hydrocarbon reserves such as those found in the Gulf states. Thus, the Gulf States started to favor workers from east, south and southeast Asian countries who were cheaper to employ. These workers, furthermore, were politically indifferent and willing to migrate without their families. The continued security precautions of the Gulf states are illustrated by Kuwait's 2011 halt in issuing entry visas for citizens of Syria, Iraq, Iran, Pakistan, Afghanistan, and Yemen. The blanket restriction was attributed to the "difficult security conditions in the five countries". The ban was partially lifted in January of 2013 (Toumi, 2013).

The Gulf states supported the popular rebellion of 2011 against the Syrian regime and strongly objected to the Party of God (Hezbollah) of Lebanon sending armed troops to fight alongside those of the regime. In the summer of 2013, GCC foreign ministers denounced Hezbollah as a terrorist group and took measures to curtail the activities of the group in member states. Consequently, many Lebanese Shia expatriates were deported, thus stoking the isolationism and vulnerability of the remaining Shia residents.

The ever-changing political environment in the region affects how ethnic, sectarian, or national groups are perceived by their governments. At the level of individual citizens, the Iranian Revolution of 1979 heightened Gulf governments' suspicion of their own Shia nationals and they started to question their loyalty. At the national scale, the Iraq War of 2003 and the subsequent fall of Saddam Hussein's regime effectively transformed Iraq from a country that had long served as a regional counterbalance for Iran to one that is a close ally to it. Iran's gradual regional ascendancy and its patron–client relationship with Hamas and Hezbollah are alarming to the Sunni-majority governments of the Arab Gulf states.

The Sunnis of the Gulf believe that their fellow Shia citizens hold their sectarian identity above their national one and are, therefore, assumed by the state to have dual loyalty at best, or to be disloyal at worst. After the Iraq War of 2003, fear of the Shia started to rise so much that the President of Egypt, Hosni Mubarak, said that the Shia "are mostly always loyal to Iran and not to the countries where they live", and before him King Abdullah of Jordan spoke of the threat posed by the emerging "Shia Crescent" in the Middle East.

The governing elite in the Arab world are almost exclusively Sunni Arabs (Syria under Assad and the post-Saddam Iraq being exceptions). They and numerous other Sunnis harbor varying degrees of suspicion of fellow Shia citizens and question their loyalty to their birthplace. This Sunni–Shia schism became more pronounced in the post-1979 Iranian revolution and became much more aggravated after the Iraq War of 2003. This perception of Shia minorities as "fifth column" was sharpened by the Iran's support for Hezbollah and vocal support for the so-called "oppressed" Shia elsewhere in the Arab world. The support from Iran and Hezbollah of the popular, primarily Shia, anti-regime uprising in Bahrain but not the primarily Sunni one in Syria has done much to paint them as sectarian, not national, liberators. This has tarnished, if not totally eroded, the anti-Western, anti-Israeli image the Iranians have projected for decades, and rekindled suspicion of the indigenous Shia in the Arab Gulf.

The demographic and sectarian issue has real security manifestations where, for example, about "1.5 million Sunnis and 1.1 million foreign workers have moved into the Eastern Province (of Saudi Arabia) and into key petroleum jobs and facilities", a fact that dilutes the concentration of the Shia in this oil-rich, strategically important area. Here, their population is "as low as 1.1 million (30 percent) of a total population of 3.9 million, and is largely concentrated outside the key petroleum facilities and cities in villages in the coastal area of Qatif" (Cordesman, 2011, p. 16, quoting Nawaf Obaid). Cordesman then adds that, "while estimates differ sharply, and lack a reliable source, it is clear that the Shi'ite population is a much smaller percentage of the total, and far less critical to the Saudi economy" (Cordesman, 2011, p. 16).

In the spring of 2011, anti-government protesters in Bahrain took to the streets and occupied a public square. Because they were overwhelmingly Shia, the government viewed them as having a sectarian, geopolitical agenda. The security forces suppressed them with violence, stoking Sunni–Shia tensions. At one point, when the protesters blockaded the financial district in Bahrain's capital city, Manama, the government became concerned about the impacts that the political crisis might have on the economic outlook of the country. This was compounded by the fact that at one point, local "security forces were overwhelmed" by the protesters (Katzman, 2011, p. 8). On March 14, 2011, and upon the request of the government of Bahrain, the GCC dispatched the joint Peninsula Shield unit security forces to assist the government of Bahrain in quelling the protests. This force included 1,200 Saudis, some equipped with tanks and armored vehicles, and 600 UAE policemen. Their task was to protect key locations and infrastructure. The land forces were bolstered by Kuwaiti naval forces who helped secure Bahrain's maritime borders. The protesters accused members of the Peninsula Shield force of taking part in suppressing them, a charge denied by Bahrain and Saudi Arabia.

In the fall of 2011, Bahrain announced that Qatar had arrested five Bahraini nationals who were reportedly planning to blow up the King Fahd Bridge, the main causeway linking Bahrain to Saudi Arabia, in addition to the Bahrain Interior Ministry and the Saudi embassy in Manama. According to a statement by the Bahrain Interior Ministry (November 2011), the apprehended men were "encouraged by others" to visit Iran and "to establish an organization in Bahrain that would carry out terror attacks on vital infrastructure and target personalities" (Momtaz, 2012).

Historically, relations between the Sunni and Shia have always oscillated between being cordial and uneasy. However, the current intense, overt, and deeply-felt suspicion and hatred that has surfaced in recent decades are unprecedented in modern history. In a critique of what he labeled the "demonization" of the Shia, Kadhim (2006) argues "there is no question that the Shia feel a certain affinity to their coreligionists in Iran. But this is a far cry from the allegation that the Shia are traitors. Sectarian affinity, however, does not capture the whole story. Our Arab brothers have done nothing to embrace us, as they keep demanding proofs of "loyalty," whatever that means." Official pronouncements of national unity aside, large numbers of Arabs generally have primary loyalty to their family, tribe, or religious leader. Therefore "secular nationalism" and loyalty to the state and its institutions are not well-developed in the Arab world. Foreign workers who are of the Shia sect, like those from Lebanon, find solace in their native coreligionists; this makes both groups suspect in the eyes of Gulf governments.

2.9 State reactions to labor grievances

Although the Gulf States have seen protests and attacks against Western places, violent terrorist incidents are rare in the GCC countries. While the US Department of State ranks Gulf countries as having a low risk of political violence, one of its recent reports warns "all residents and visitors" to Qatar to remain "alert and aware of the significant continuing threat from terrorism in the region. Terrorist attacks could be indiscriminate, occur with little or no warning, and be conducted against ... Qatari interests" (US Department of State, 2012a). Between 2011 and 2012, Bahrain, Oman, and Saudi Arabia experienced varying degrees of public protests. Bahrain's protests ebbed and flowed ranging from large and sometimes violent events to small yet noisy affairs.[4] The continued instability, low level as it may be, could have spiraled out of control. Matthiesen (2013, p. 84) contends that in the "medium-term it cannot be ruled out that Western expatriates, particularly British citizens, might be targeted for their governments' alliance with the Bahraini royal family". If a scenario like this were to materialize, it could push thousands of professional expatriate workers to leave the country. This would jeopardize daily operations of the technicalogically complex infrastructure such as that for water and energy, gradually degrading water security in the country.[5]

---

[4] Anti-regime activities in Bahrain were simmering well into 2013, when "youths armed with petrol bombs attack police on an almost nightly basis in villages and towns around the capital" (BBC, 2013).

[5] Since the early to mid 2000s, the Gulf states started building strategic water reserves which could be tapped in cases of emergencies, which may be triggered by human or natural forces. Although Kuwait had been a pioneer in this area, its reserves were said to meet the people's needs for 7 days in 2000.

Although Oman had smaller protests, these managed to disrupt businesses operations. The political and security climate in Oman is under control such that Westerners are as safe as the natives. The US Department of State considers the country to be such a safe-haven that it reported that "[t]raffic accidents remain the biggest threat against Americans in Oman" (US Department of State, 2012b). During the winter and spring of 2011, workers engaged in large-scale strikes involving both private and public sectors. Protesters blocked roads and set a government building aflame resulting in injuries and a fatality. Because the people of Oman continue to strongly support and respect Sultan Qaboos, protesters limited their demands "to calls for additional jobs, higher wages, and a more responsive government" but did not call for radical political change (US Department of State, 2012b).

While the Gulf states respond differently to labor protests and unrest, their overall approach is one of appeasement and firmness. The UAE government, for example, started requiring "employers pay for migrant workers' travel, employ-ment permits, medical tests, and health care. It has also closed down some workers' camps that do not meet health and safety standards" (Agence France Presse, 2008a). Expatriate workers and their dependents who live in Abu Dhabi are required by law to have health-insurance coverage. The Abu Dhabi Health Insur-ance Regulation Bill under Law 23 states that employers and sponsors must provide health insurance for expatriates who live or work in the emirate of Abu Dhabi, even if their visa is issued in a different emirate (El Shammaa, 2008).

In 2010, the Municipality of Dubai started demanding that companies with offices within its jurisdiction house their workers in the emirate itself – not in neighboring ones. Some of the old and dirty labor camps were demolished or upgraded and many companies relocated their workers from "cramped labour accommodation in Sharjah and Ajman ... (and) companies can save three hours of travel time and can improve workers' productivity" (Sathish, 2010). Dubai's Municipality and Ministry of Labor guidelines require that a room with an area of 240 square feet should accommodate a maximum of six people or three double bunk beds and "at least one bathroom for every eight people" (Sathish, 2010). Until then, authorities had turned a blind eye when companies were "forced" to wedge up to ten workers per room due to housing shortages (Sathish, 2010). The silence over the need to provide upkeep for the housing facility, the cramped residential conditions, and the cursory enforcement of regulations raises questions about whether workers will feel much difference in their quality of life.

Setting a minimum wage reduces the exploitation of workers by preventing employers from cutting wages below a certain level. However, it requires employers to bring wages to around $300 per month, a move that prominent business people and those with connections to ruling families will resist vigorously. Moving forward, setting a minimum wage not only satisfies expatriate demands but it will also improve bilateral relations with some foreign governments. For example, the Philippine Department of Labor and Employment and the Philippine Overseas Labor Office assert that their laws require domestic workers who are deployed abroad to receive a minimum monthly wage of $400. Politically connected corporate leaders in the Gulf states and hiring agencies resist the idea of a minimum wage.

2.10 Targeting Western labor

Political turmoil in one country carries the prospect of a transboundary spillover effect into geopolitical neighbors. The prolonged instability in Yemen has been a concern to the Gulf states, especially Saudi Arabia. Since the mid 1980s, the Gulf states have viewed Yemen through the prism of its position on Iraq's occupation of Kuwait, the presence of senior Al Qaeda members there, the Houthi rebellion in the north, and the smuggling of illicit goods and people (e.g. Yemeni and African immigrants) to Saudi Arabia and other Gulf states. The dormant Houthi rebellion became active in 2004 and was fanned further after the 2011–2012 popular revolt against President Saleh. The Gulf states are increasingly concerned with the Yemeni government's growing inability to assert control over large swaths of territory in the south (Al Qaeda) and north (Houthi) of the country, and the spillover potential that might have. Saudi intelligence services are active in Yemen, where they are believed to have penetrated Al Qaeda in the Arabian Peninsula. Their work contributes to peace and stability in Yemen and the rest of the Arabian Peninsula.

Weapons, ammunition, and drugs are regularly smuggled from Yemen into Saudi Arabia, and some make their way into other Gulf countries. The threat posed by Al Qaeda in the Arabian Peninsula against Saudi Arabia spiraled in the fall of 2003 and reached a peak in the summer of 2004. In 2003, Saudi authorities began a strong crackdown and cooperated with the United States and other countries to investigate, track, and gradually neutralize this threat to their national security. Terror attacks targeting westerners struck Paul M. Johnson Jr., an American defense contractor. Members of Al Qaeda beheaded him and posted a video of the killing on the Internet (Whitlock, 2004). In December of 2004, members of Al Qaeda launched an armed attack on the US consulate in Jeddah, Saudi Arabia, killing nine and injuring many. The barbaric nature of the former act, and the daring, brazen midday raid on the heavily fortified consulate compound – which resulted in an extended exchange of gunfire, explosions, and pillars of smoke billowing from the site – coalesced to frighten some essential and many non-essential employees into leaving the kingdom. In 2004, it was estimated that 38,000 westerners left the country; they were mostly white-collar workers employed in a variety of technical and management positions in the oil, education, banking and finance, and medical sectors (Ambah, 2004). These workers make up the backbone of the national economy. Hence their sudden departure dealt a blow to the country. Additionally, some non-essential personnel and families of American diplomats were ordered to return home (Ambah, 2004).

Many Western expatriates have been in the Gulf for years and have learned to accept that a certain level of erratic experiences come with working in the region where some countries experience significant levels of turmoil, and the dominant culture and system of government are different than those in their home countries. They have also known the GCC countries to be among the safest in the Middle East. Irrespective of these facts, the security environment has the potential to change dramatically, and in a relatively short time. Manifestations of this were evident in the uprisings of the Arab Spring in Syria, Yemen, and Egypt. Furthermore, based on a survey of 32,000 professionals from 140 countries working in the GCC countries, GulfTalent, a recruitment agency, noted that the mass anti-government protests in some Arab countries, which were beamed around the world, "may deter some Western professionals from relocating to the Gulf" (Reuters, 2011b). This, and the additional security precautions that Gulf governments took after the Arab Spring, made it harder for corporations to find qualified people for certain positions such as professors of science and engineering.

2.11 Protests and promises: more of the same

The occasional protests by expatriate workers are related to one or more of the following triggers: rising cost of living (2005–2007), low pay, poor housing conditions, and inadequate transportation services to job sites. The first signs of unrest in the UAE were visible in the fall of 2004, when thousands of Asian workers marched down the eight-lane Sheikh Zayed Highway toward the Ministry of Labor (Davis, 2006). In 2006, a protest by some 2,500 workers in one construction site in the UAE sparked a night of violence which caused an estimated Dh 3.5 million (over $1 million) in damages. A year later, a group of predominantly Indian construction workers in Dubai, who were protesting against harsh working conditions, vandalized police vehicles and public property. In 2008, the Kuwaiti police used batons and teargas to break up a protest by thousands of Bangladeshi workers who were demanding a significant pay raise and better living conditions. An estimated 240,000 Bangladeshis work in Kuwait where a large number of them are employed with cleaning companies; their salaries were said to be $75 per month (Cardozo, 2008). In the following year, some 7,000 workers in Bahrain went on strike demanding higher wages; they were receiving a base salary of $185 per month (Millington, 2009). In 2008, as food prices were soaring, hundreds of expatriate workers rioted in the emirate of Sharjah, demanding higher wages. While wages varied among companies, some construction workers were receiving less than $170 per month. They burned offices of a US-owned contracting firm, torched cars, and damaged buses (Cummins, 2008). Angry workers also "tried to assault" policemen and labor ministry officials who were at their housing compound (Agence France-Presse, 2008a). Overall, what makes the demonstrations in the Gulf states noteworthy is their scale, intensity, relative frequency, and the fact that powerless, often illiterate, workers dared to make their voices heard despite the fact they were breaking the law, which bans all forms of protests.

Remittances have a real effect on the financial health of individual families and impact the national economy. The volume of remittance from the GCC countries went from $30 billion in 2007 to $40 billion in 2008. Emigrant Bangladeshis around the world remitted an estimated $10.98 billion to their home country in 2009–2010 and Bangladeshis in the UAE remitted more than $1.89 billion in 2010 (Forsythe, 2011). Currency fluctuations affect foreign workers' remittance levels, which impact their family members who typically stay in their country of birth. For decades, GCC countries have pegged their currencies to the American dollar except for Kuwait, which decided in 2007 to link its dinar to a basket of international currencies with different weightings. When the value of the dollar slipped considerably in the mid to late 2000s, it had an adverse impact on expatriates in the Gulf. When foreign workers converted their salaries to their respective national currencies, they received smaller amounts, which made them agitated.

A common consequence of a labor protest is to identify its leaders and send them back to their countries of origin. In 2008, Saudi authorities evicted 40,000 Bangladeshis for "provoking labor unrest, strikes, and protesting low wages and poor labor conditions" (US Department of State, 2011). The scale of this eviction is higher than usual and may reflect a policy of spreading fear among foreign workers so as to deter similar actions by others. Another indicative anomaly is that the Gulf states have a large a number of undocumented workers, and their infractions range from those who entered illegally to legal migrants who did something that nullified their visa status. As part of Saudi Arabia's routine crack down on illegal immigrants, authorities arrested more than 150,000 visa violators in the metro Jeddah area in 2005 (Shah, 2006).

In June, 2007, undocumented and illegal workers hoping to leave the UAE were offered free one-way plane tickets to their home countries, with no questions asked. Emirati authorities were "swamped by 280,000 workers who, fed up with the rising cost of living and low wages, were ready to go home" (Surk, 2007). Since the 1990s, the Indian economy has been experiencing high growth rates and generating a large number of new jobs, which makes it harder for some to justify leaving the homeland for a relatively low-paying job in a foreign culture.

Since the mid 1990s, governments of the Gulf states have periodically declared a general amnesty for illegal immigrants and those who have overstayed. In 2003, an amnesty program enticed some 100,000 persons to leave the UAE. On the other hand, the context of illegal immigration in Saudi Arabia is unique, and its scale is much larger; the number of yearly deportees is estimated at 700,000 (Shah, 2006). Many of these people have entered the country for Umrah or Hajj but did not leave when their visa expired. In early 2000s, the Kuwaiti government estimates put the number of expatriates who have violated their visas at 60,000 (Shah, 2006). In 2013, in Saudi Arabia, about half of the eight million foreign workers were believed to have improper papers and were hence working illegally. It is worth noting that many foreigners stay in the Gulf for over ten years and some for decades, which naturally leads them to develop a certain level of affinity to their place of residence.

Sometimes, Gulf governments deliberately tighten the screws on foreign workers. Kuwait is considering excluding them from subsidized services like water and electricity, a change that would cut deeply into the already meager wages of unskilled workers. Qatar addressed its severe traffic congestion partly by discontinuing the issuance of driving licenses to unskilled expatriates (this classification included secretaries, computer technicians, cashiers, cooks, and tailors). This measure would "imply that only Qataris and foreign professional workers – estimated together to comprise roughly 25 percent of the country's population – will be the only ones left on the road, leaving less skilled workers dependent on almost non-existent public transportation or their employers to arrange for them to get to and from work each day" (Coker, 2013).

Governments of the Gulf are sometimes blamed for continued labor unrest because they do not seriously address the persistent grievances of workers, which explains the recurrence of demonstrations. In Kuwait, for example, the government's response to striking immigrant workers by introducing minimum wages for some sectors, so, for example, cleaning staff now earn a minimum of KD 40 ($140) and security guards KD 70 ($600). It also started limiting the number of immigrants of certain nationalities (Kuwait Times, 2008). The latter, in fact, has been quietly implemented in most Gulf states where decision makers moved to dilute the concentration of expatriates who hail from the same country. The mass protests that Gulf countries started experiencing in the early 2000s led some governments to reduce the number of workers from particular nationalities in certain sectors of the economy. The Gulf governments are concerned about a group of co-nationalists coordinating a labor action that would paralyze a critical service, possibly holding the government "hostage" to their demands. Hence they have been taking measures to diversify the employment mix in services like first respondents (ambulatory service), police, and on other related sectors.

The very firm response of Saudi security agencies to the attacks of 2003–2004 helped contain the threat, mitigate its impacts, and assure expatriates so that the country's economy was saved from a possible mass exodus. It is worth noting that in light of the attacks, the American ambassador to Saudi Arabia told US nationals to go back home for their own safety (Associated Press, 2004; US Department of State, 2004). The Gulf states took notice of this security challenge and initiated preemptive measures such as appointing qualified nationals to strategic positions that would ensure business continuity of critical operations in some of the key industries, particularly in the energy sector. For example, highly trained petroleum and chemical engineers in some Gulf states were moved from academic posts and given senior appointments in national energy companies. Human-capital planning became part of the strategic and national security calculus of the oil and gas sectors, primarily because there are insufficient numbers of skilled nationals. Consequently, normative labor nationalization, wherein locals replace foreigners, is an unrealistic goal. Because smaller Gulf countries such as Oman, Bahrain, and the UAE don't have a strong security network like that of Saudi Arabia, they stand vulnerable in the face of sustained internal or external challenge. The effects of labor-triggered instability on the water and energy sectors cannot be underestimated.

The Gulf governments' usual responses to labor strikes, demonstrations, and occasional concomitant acts of violence are to make minor concessions while also issuing draconian threats against participants. They deport suspected leaders and organizers of such labor action, but they also acquiesce to some of the demands that are usually related to low, and sometimes unpaid, wages, unsafe working conditions, and to poor living conditions such as a chronic shortage of water for cooking and bathing at labor camps. Such a carrot and stick response is intended to neutralize labor unrest. However, when the immediate challenge comes under control, decision makers sometimes do not introduce reforms that were promised to striking workers or, more commonly, they fail to deploy a sufficient number of inspectors in the field to ensure that companies are abiding by the newly introduced laws. Consequently, companies often return to flouting local, and sometimes international, labor laws with respect to paying suitable wages and providing appropriate accommodation, so labor unrest is reignited.

While some Gulf states have historically acknowledged the need to reduce the number of foreign workers, no country has taken active steps in that direction, save for the policies that were adopted by Kuwait and Saudi Arabia in 2013. For example, Kuwait decided to reduce the number of foreigners by one million within a ten-year framework. Shortly after the policy was announced, authorities started taking steps to reach their target of 100,000 evictions target for that year. Saudi Arabia's goal is much larger. These major and recent policy changes show that social policies in the Gulf states are in a state of great flux and uncertainty.

2.12 Cyber attacks

A report by the US Geological Survey, an American governmental agency, stated that the primary threats to water security include terrorism, population growth, arid climatic conditions, cyber and industrial sabotage, and infrastructure breakdown (Tindall and Campbell, 2010, p. 5). Virtual attacks on classified military computer networks, uranium centrifuges, oil and gas pipelines, oil fields, and water infrastructure have been on the rise. These technological systems are the backbone of modern economies and urban living. If they were to fail for an extended period, their socioeconomic impacts would be catastrophic. An example of just how vulnerable society's water infrastructure is comes from a recent report (Dilanian, 2011) that a large southern California water system hired hackers who, in a single day, simulated their ability to commandeer "the equipment that added chemical treatments to drinking water", which would have made the water undrinkable. It concluded that the United States is "vulnerable to a cyber attack, with its electrical grids, pipelines, chemical plants and other infrastructure designed without security in mind". Water infrastructure. which includes distribution systems, reservoirs, and pumping stations, are a potential target for terrorist attacks because freshwater is central to human existence and economic operations. Water infrastructure is also linked to many other services (e.g. electrical power) and is often considered a soft target. The psychological reverberance of disrupting water supply to a major metropolitan center would be immense, a key goal of terrorists or warring factions. Finally, it is worth noting that water infrastructure has been targeted by antagonists throughout history (Gleick. n.d).

The Saudi state-owned oil company Aramco had it computers attacked by a virus in "what is regarded as among the most destructive acts of computer sabotage on a company to date". The attackers, who tend to seek maps of new oil fields or research on efficient photovoltaic generators, were able to erase documents, spreadsheets, e-mails and other data on 75 percent of the company's computers, and display the picture of an American flag in flames (Perlroth, 2012; Lipscombe, 2014). The so-called Shamoon virus caused $15 million dollars in damages; its bigger impact was the "shockwaves" that it sent "through an industry that had been previously immune to large-scale data security breaches" (Lipscombe, 2014). Subsequently to this attack, the computers of RasGas, Qatar's liquified natural gas provider, were struck with a similar virus. The US Secretary of Defense Leon Panetta said that "All told, the Shamoon virus was probably the most destructive attack the private sector has seen to date" (Williams, 2012).

Analyses of the attacks' "footprints" reveal that the perpetrator is likely "a company insider, or insiders, with privileged access to Aramco's network" (Perlroth, 2012). The company has more than 55,000 employees of whom 70 percent are Shia Muslims, and some hold critical engineers positions. The Saudi and American governments believe that the primary suspect is Iran. A report by a Brookings Institute scholar states that "The Saudi Ministry of the Interior has long been obsessed with concerns about Iranian intelligence activity among the Shia minority. The Saudi Shias are not Iranian pawns, they have very legitimate grievances just like the Bahraini Shia do, but a few will take Iranian help and training" (Riedel, 2012). It only takes a few individuals who are sufficiently disgruntled to wreak significant economic havoc on vital infrastructure. The Shia of Saudi Arabia and of other Gulf states are distrusted by the Sunni majority (or ruling class) and by the government security services. They feel that they are not well represented in government, and that their religious freedom is curtailed. A report to the American Congress asserts that the UAE "has long feared that the large Iranian-origin community in Dubai Emirate … could pose a 'fifth column' threat to UAE stability" (Katzman, 2010).

Saudi Arabia's Ministry of the Interior has always believed a Shia terror group with links to Iran was responsible for the attack. A Brookings Institution, Bruce Riedel (2012) concludes that "Aramco, in short, is a target-rich environment for angry Saudi Shias with ties to Iran". The next war in the Middle East may be a stealth cyber war whose victims are just as real as any actual war.

2.13 Conclusions

Universally speaking, threats to water supply fall into three broad categories: manmade, natural, and technological (Table 2.3). The Gulf states' concern over guest workers has been the subject of intense discussions in the last few decades.

Table 2.3 *Common threats to water security*[*]

Manmade (anthropogenic)	Natural	Technological
Terrorism [#]	Climate change	Infrastructure failure
War and civil unrest	Hurricanes	Hazardous chemicals and
Population growth	Earthquakes	biological material events
Human error and poor assessment and resource allocation	Tsunamis	Malfunctions of information
	Droughts	technology and equipment
	Floods	
	Wildfires	
	Landslides	
	Volcanoes	

[*] Source: Tindall and Campbell (2009).
[#] Includes foreign and domestic cyber and industrial sabotage, particularly against/ including Supervisory Control and Data Acquisition (SCADA) control systems.

Guest workers have become more assertive as evidenced in the more frequent demonstrations. The first Gulf War of 1991 confirmed the perception among many natives that their reliance on foreign workers is a national security threat. To paraphrase the Swiss author Max Frisch (as quoted by Matthews, 2013), the Gulf states asked for workers but received human beings with all the complexities that come with foreigners "temporarily" settling in a country and culture that are very different from their own.

While the Gulf states appear to have survived the Arab Spring almost unscathed, some seemingly common events like labor unrest, or the frequent parliament-related turmoil in Kuwait or demonstrations by the Shia communities in Saudi Arabia or Bahrain, may spiral out of control.[6] The Sunnis view these communities as having allegiance to Iran and to Shia clerics and hence a distrusted "fifth column." Native Shia citizens have been arrested and charged with spying for Iran or with planning acts of terrorism. Cyber terrorism is increasingly becoming an option to frustrated workers or oppressed religious minorities. Targeting water or energy infrastructures is a real possibility in the socially and politically unstable Middle East.

Some countries appear to accept the reality of long-term dependence on foreign workers, and have taken some steps to improve the quality of life for these workers. After Qatar was granted the right to host the World Cup in 2022, its mal-treatment of low-skilled foreign workers became a rallying cry of various groups and media outlets around the world. This pressure appears to have led the small Gulf state to permit foreign workers to form their own sports clubs in 2013. It also initiated the construction of West End Park, a new entertainment complex meant for exclusive use of the country's low-income laborers. The complex will house a hypermarket, theater, shopping mall, movie theaters, and a cricket stadium. The project will have a residential area with some 80,000 residential units. The complex will host events affordable to foreign workers where their cultural traditions are celebrated such as Indian National Day and Chinese New Year (Akkad, 2013). While such steps are in the right direction, a lot more needs to occur if Gulf states are to diffuse the frustrations of workers and protect their national and water security. They need to facilitate the acculturation or socialization of expatriates, and develop policies that will build mutual confidence between

---

[6] A worst case scenario would be for a central government to lose control over a peripheral part of the country to frustrated and marginalized minorities, or to foreign workers. A short-term loss of control would send a negative signal to residents and investors. A situation of prolonged instability would radicalize these elements and might attract transnational extremists, some battle-hardened, from the wider region. It would produce a shock to the national economy, deter foreign investment, and may disrupt operations of critical infrastructure including facilities that produce desalinated water or treat wastewater. Given the Gulf governments' seemingly successful military intervention in Bahrain, and their generous economic measures towards their citizens, this seems an improbable scenario.

native and foreign residents as these will "result in a reduction of the potential threat posed by these expatriates to Gulf political stability", and will also increase their satisfaction, which in turn boosts their productivity (Atiyyah, 1996). Early in 2013, the government of the UAE initiated a major information and education campaign intended to help Emiratis and expatriates better understand each other. It seeks to enhance the locals' and foreign workers' awareness of national identity and to increase their awareness of the country's history, culture, and tradition. It involves schools, university lectures, and exhibitions. The general director of the government agency responsible for the program (Watani) linked the general understanding of national identity with communities living in "greater harmony (Radan, 2013; Watani.ae)."

Many of the foreign workers in the Gulf states are effectively becoming immigrants, even as governments do not officially recognize that. They are finding ways to stay for decades, often well past their employment status expires, and to have family members join them in their adopted country (Lori, 2012). A significantly improved treatment of guest workers may well make them feel "at home" and rooted in the host country; this would make it harder to send them back to their countries of origin. Would this group peacefully accept their eventual deportation?

The responses of Ethiopian workers who were deported by Saudi authorities sometimes spiraled down to street violence because some resisted orders (Abebe, 2014). This was partly because many of them had been in the country for a very long time, and viewed themselves as Saudis. On the other hand, how would the indigenous population react to creeping naturalization of a select number of workers whose cultural values are very different from local ones? These seem to be intractable conundrums, but must be faced by the Gulf states in the near future. A pathway to permanent residency whereby select expatriates who meet certain reasonable criteria would be allowed to apply for citizenship could act as a mechanism to vent their frustrations, and it would boost the population base of the small Gulf states.

Despite some improvements in living and working conditions of foreign workers, these changes affect a very small number. That is to say, their grievances and those of the Shia population simmer unresolved and occasionally manifest themselves in street protests. If a small fraction of these communities decided to express their frustrations by resorting to violence, they might target the host countries' critical infrastructure – be it oil or water. A massive successful attack on either one could translate to a severe disruption of water supply with all manner of potential political fallouts. Furthermore, desalination technology, like all complex technological systems, is prone to fail. Given the massive dependence of urban centers in the Gulf states on desalinated water, the authorities must remain vigilant in an attempt to prevent a catastrophic failure of a major plant supplying water to a large city like Jeddah or Riyadh. Only when nature or people put this scenario to the test will we know how prepared these countries are to respond to a large-scale threat to their critical infrastructure, and what the security effects of a post-system failure will be.

3

Outsourcing farming

3.1 Introduction

Over six decades ago, an academic report concluded that the future of Saudi Arabia "lies in the wise utilization of its most precious but limited resource, and in the methods of safely capitalizing on it" (Crary, 1951, p. 383). This observation about the central role that freshwater and its management play in the arid kingdom proved to be timeless. Food in the Gulf states used to come from farming nearby lands, and from fishing for coastal populations. They also imported small amounts of specialized food like rice that eventually became a national staple.

Sustainable water development and planning needs to consider biophysical limits, and socioeconomic realties. It also needs to pursue multipurpose development goals simultaneously in a coordinated manner that advances human welfare within the constraints of the ecosystem. Dwindling groundwater supplies at the home front, the global surge in food prices in 2007–2008, and the imposition of export restrictions on grain by major producers such as India, Vietnam, and Cambodia have raised the concerns of many Gulf states and led them to pursue creative approaches to food security, which included outsourcing some of their agricultural activities.[1] Economically advantaged countries who are also food deficient have been acquiring farmland, often in countries that are poor cum weak, and run by corrupt and autocratic leaders, an approach that has led to charges of "land grabbing." The essence of this challenge is captured by John Maynard Keynes' comment almost a century ago; he wrote that "the political problem of mankind is to combine three things: economic efficiency, social justice and individual liberty" (Kuttner, 1987, p. 1). As the Gulf states pursue food

---

[1] Qatar elected to pursue limited food self-sufficiency within its secure borders by embracing technology-centric farming.

security under ecologically efficient conditions, they are being confronted by charges of dispossessing native populations and squandering their land rights.

Food supply is affected by extreme weather systems, regional or international political considerations, or by upheavals that disrupt trade routes. These events may be short- or long-lived, and could affect one or multiple crops. The United States is normally the largest grain exporter followed by Australia; the latter produces around 25 million tons per year. Australia's two record-breaking droughts in the first decade of the twenty-first century had crippling effects on its water supply and food production. This reduced its 2006 grain harvest from around 25 million tons to merely 9.8 million tons, and boosted prices on the international market (Bryant, 2008; see also Barry, 2008). Furthermore, a poor wheat harvest and soaring rice prices in 2007 led the Indian government to impose a ban on grain export, and to argue that its duty was to its own poor. Subsequently, it decided in April 2008 to make a concession for Bangladesh, its neighbor and largest export market. By selling them 500,000 tons of rice "at prices less than half of those prevailing in international markets at the time … probably averted a humanitarian disaster in Bangladesh" (Headey and Fan, 2010). A report by Oxfam (Welton, 2011) states that "there is a strong inclination for exporters to impose export bans in reaction to potential food price increases in their own country."[2] This is the political and economic context that worried the Gulf states who promptly decided to improve food security for their respective countries.

Export restrictions disrupted grain supply and sensitized people and decision makers to the virtual water embedded in food products, and ushered a tighter conceptual nexus between water and food security. Similarly, while river systems visibly and physically connect riparians who share a transboundary watercourse, international food trade masks a real, but invisible, hydrological link that binds most countries of the world. Food-deficient countries are connected to a global supply chain that provides them with virtual water that they consume as food products. Trade and technology facilitate the movement of huge amounts of embedded water to the Gulf states. These security-sensitive actors who are also "virtual downstream states" in the foodshed were motivated to expand the hydrological reach of their "food domain" by diversifying their food security portfolio. While the watershed is a natural system, the foodshed is a social–natural hybrid (Feagan, 2007) system that conjures images of "streams" of food flowing towards a focal point. We will examine Arab food and water security through the lens of foodsheds in this chapter.

---

[2] The report finds that such strategies "exacerbate problems … and may damage incentives to increase production at home longterm" (Welton, 2011).

3.2 Globalizing the local foodshed

Edward Thompson (1971, pp. 76–77) described the food protests in eighteenth-century England not as mob or riot activities but as "rebellions of the belly" or reactions to economic adversity. He argues that food uprisings were expressions of deep anger due to the violation of an implicit social contract (pp. 78–79). These observations have relevance for the contemporary Arab world where, in the 1970s and 1980s, Egyptians and Jordanians rose up against the lifting bread subsidies (Cowell, 1989). More recently, the general spike in food prices where some staples doubled in price (Economist, 2012) triggered protests in numerous developing countries, especially in Africa. In Egypt, for example, domestic food prices rose 37 percent between 2008 and 2010 (Economist, 2012). Some of the protests became violent, and scores of people were killed. In addition to the important role food prices played, other economic and political issues contributed to the charged atmosphere in which protests were started (Berazneva and Lee, 2013). The relatively sudden and sharp rise in food prices in 2008 triggered what started out as "bread riots" in Bahrain, Yemen, Jordan, Egypt, and Morocco and turned into the Arab Spring that toppled some regimes (Economist, 2012; Perez and ClimateWire, 2013).

Significant growth in population size and in per capita food consumption, along with geopolitical instability in the Middle East, have led many governments and investors to expand their resource capture and take on greater responsibilities to ensure food security for the people. Furthermore, a report by the National Commercial Bank of Saudi Arabia stated that food forms between 15 percent and 30 percent of the consumer price index in the Gulf states. The exchange-rate peg had limited the abilities of central banks in terms of doing much to mitigate inflationary pressures. The report concluded that food inflation "presents a potentially considerable socioeconomic risk which the authorities are poorly equipped to deal with" (Kotilaine, 2010).

While buying food on the international market is always an option, the politicization of food trade by some countries has tipped Arab leaders to want to have greater control over the food supply chain. For example, between 2006 and the middle of 2010, Israel imposed a strict blockade against Palestinians living in the Gaza Strip. Gisha, an Israeli advocacy group, described the blockade as tantamount to "economic warfare" and collective punishment against the 1.5 million residents of Gaza. Israeli documents reveal that during that period its military "made precise calculations of Gaza's daily calorie needs to avoid malnutrition", and sought to "keep Gaza's economy on the brink of collapse". For the Israeli army, the purpose of the blockade was to weaken Hamas, its archenemy (Associated Press, 2012). Israel calibrated its economic siege of Palestinian territory by withholding a lot of the food supply, which caused real suffering yet allowed enough food through so as not to cause a dire economic crisis that might generate images of "hungry children". Dov Weissglas, an advisor to the Israeli prime minister Ariel Sharon, joked "it's like a meeting with a dietician. We have to make them [Palestinians] much thinner, but not enough to die" (Benn, 2006; see also Levy, 2006).

In 1798, Robert Malthus voiced a belief, later echoed by more contemporary individuals and organizations such as the Club of Rome, that, given rapid population growth, humans are likely to reach the carrying capacity of natural resources; when they do, starvation and death will prevail. Esther Boserup reversed the long-established logic by arguing that population or resource pressure would spur innovation and greater reliance on technology and better management approaches that would help in boosting production in line with the rising demand (Boserup, 1961; Netting, 1993). Improvements in the transportation and communication technologies and the end of the Cold War have allowed Boserup's historical physical–social continuum to be extended

Homer-Dixon (1999) argues that rapid population growth and waning renewable natural resources lead to resource capture by powerful actors who use their influence to skew resource distribution in their favor; they do this by leasing, purchasing, or seizing lands that are beyond their control or national boundaries. Using a political economy perspective to scarcity, Ruckstuhl (2009, p. 5) argues that "renewable natural resources define systems of power and access" through the ownership of a resource, how it should distributed, used, and finally how it is governed. However, the asymmetry in political and economic power between actors make the weaker one vulnerable as it needs to appease or consider the interests of, say, a major foreign investor before making certain decisions that may affect its interests. This is one of the critiques of foreign investments in agricultural lands abroad. Resource pressures, concentration of wealth, ever larger scales of production, and advances in technologies are recasting established relationships between people and natural systems (Homer-Dixon, 1999), and changing the functions and meaning of "place."

Food which consumes immense volumes of water to grow is increasingly transported long distances around the world until it reaches its focal points of distribution and consumption. Similar to the precipitation that passes through drainage systems to carve the Earth's surface into watersheds, modern production and long-distance movement of crops are enlarging our understanding of "foodsheds." However, unlike the metaphorical meaning of a "natural" watershed, people in the globalized foodsheds are reduced to consumers, and ecologically diverse ecosystems to mechanized monocropped landscapes. The net effect is greater physical and ethical detachments of consumers from the natural and human environments that produce their food.

Over the centuries, food acquisition has been transformed from hunting and gathering to farming small plots that are adjacent to one's residence and attended to by family members, to one whose locus is further away, where domesticated animals serve farming, transportation, and dietary needs. The advent of the industrial revolution led to the mechanization of agriculture, which increased the size of the farm while reducing the number of people needed to work the land, and led to the eventual introduction of large-scale commercial farms. There was an evolution in the relationship between a person and his/her community which changed "the place" of farming, socially and physically. Similarly, settled agriculture, mechanization, and a higher quality of life has made peoples' needs and food supply systems more complex, affecting the conceptualization of where the locus of responsibility for food security lies. It has moved from the personal to the family level, then to the rural community level, and eventually central governments have started to play a larger role. The form of government practiced in the Gulf states, where leadership is inherited and the head of state presents himself as a "father figure," implying a greater responsibility to the people, and providing for their food security.[3] They, along with some large corporations, have purchased or leased lands abroad that they intend to grow food on, and bring most of the produce home. This global expansion of their foodsheds is a natural evolution of demographic and economic developments at the local level, which has intensified the demand for water (Figure 3.1). As the pressure on the foodshed in the Gulf states has made them increasingly incapable of supporting the needs of residents, the capacity of national governments to fill in the shortfall has grown at a faster rate. Furthermore, the emerging globalized foodshed, especially international investments in farmlands, is likely to have geopolitical effects that could influence international relations in ways that are new and challenging.

People who work the land develop a deep attachment to their source of sustenance and existence. This attachment is particularly strong in Arab culture, where land is associated with one's personal honor. An Arabic proverb states that "your land; (is) your honor" (*Ardak A'rthak*). Globalization is, however, relaxing the intensity of these feelings and Arabs' attachment to lands that have been passed over from generation to generation. Placelessness is about "the weakening of distinctive diverse experiences and identities of places" (Relph, 1976, p. 6). In countries like Egypt, Lebanon, and Yemen, emigration is a routine social experience; the "new normal" for communities that are politically troubled, and/or economically poor. Just as the industrial revolution removed small artisans and manufacturers from households and neighborhoods, the current pressures on land

---

[3] The native population of the Gulf states enjoys extensive cradle-to-grave social safety nets and generous subsidies to many services; more than what most other wealthy governments do for their citizens.

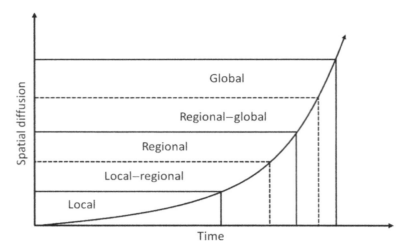

Figure 3.1 Diffusion of food production and acquisition

and water, compounded by wealth accumulation and urbanization, are pushing food production farther and farther away from the market place and the dining table. This growing estrangement between consumer and producer, where methods used in the production, processing, and packaging, raise questions about the nutritional security of imported food (Langelaan, 2013).

Political boundaries and national laws are human constructs that have tangible implications for residents, yet they are fleeting notions to global investors or those with access to the levers of power. The forces of globalization and technological change are contributing to the transformation of distinctive places and the dilution of national identity by homogenizing local experiences that affect the meaning of space and individual sentiments that are traditionally associated with the home-land. At the individual level, the political ramifications of these processes may include the weakening of indigenous nationalism and the emergence of a new transnational identity. At the level of international trade, corporations are guided by meta-values such as economic efficiencies and the security of their supply chains. Meanwhile, most people are reduced to passive spectators regarding the ecological conditions where their food is grown, and the political and economic circumstances of how it is produced, handled, stored, and transported. These global dynamics are also changing the meaning of home whereby the political, "senti-mental home" will be static and legally circumscribed while the boundaries of the "economic home" are being expanded and redefined.

"Foodshed" was first used as a tool to help map out the flow of food from the farm to the supermarket, and many used it to refer to the flow of food from local producers to consumers. Hence, it has been used to describe alternative food production systems, and their social and environmental impacts (Feagan, 2007; Peters *et al.*, 2009). Foodshed analysis is about actual or potential sources of food for a population (Peters *et al.*, 2009; Hedden, 1929; Kloppenburg *et al.*,1996) and the risks associated with transporting this resource to the marketplace. In this chapter, foodshed is understood as a unifying metaphor that refers to the geographic areas from which foodstuffs are transported to markets and consumers.

3.3 Globalizing food security

In an effort to bolster food security, national governments have focused on production within their national boundaries. In the Gulf states, land reserves with farming potential have been converted to arable land since the 1970s (Table 3.1). Most lands in the Gulf states are characterized by soil profiles that are almost exclusively (80 to 90 percent) sandy in texture with good drainage, very low water retention capacity, almost absent organic content, and therefore deficient in nutritional materials needed for plant growth. Many of the sandy soils have a moderate to strong alkaline pH level. The exceptions to this broad picture are the mountains of Oman that occupy 15 percent of the total area of the country. Oman's Dhofar Mountains in the extreme south rise up to 2,500 meters in elevation and are rather humid. The southeastern corner of Saudi Arabia and the northeastern part of the United Arab Emirates (UAE) and of Oman receive above average levels of precipitation (FAO, 2008). Irrigation is integral to the process of making land

Table 3.1 *Overview of the agricultural sector in the GCC countries*

	Labor force in agriculture (percentage of total labor force)		Arable land (as percentage of land area)*		Land under irrigation (area, in thousands of hectares)			Water use in agriculture (percent of total)
	1996	2011	1980	2012	1985**	1995**	2009	2007
Bahrain	1.53	0.61	2.9	2.1	1	4	4	45
Oman	40	28	0.1	0.1	41	62	59	88.4
Kuwait	1.13	0.99	0.1	0.6	2	5	11	54
Qatar	1.82	0.66	0.3	1.1	5	13	13	59
Saudi Arabia	13	5	0.9	1.5	1,150	1,620	1,731	88
UAE	6	2.98	0.2	0.6	58	68	230	83

Source: FAOstat (n.d.).
* World Bank Data (n.d.).
** Bazza (2005).

arable and more productive. Since the 1960s, global crop production has increased by three times due to higher yields per unit of land, farming intensification that was driven by multiple cropping and shorter fallow periods, and, to a lesser extent, by the expansion of arable land area. During this period, the area of arable land decreased in developed countries and increased in developing ones. In recent decades, most of the the Gulf states accelerated the development of their agricultural lands in the early 1980s. For example, the size of arable land in the UAE went from 0.2 percent of the country's land area in 1980 to 0.6 percent in 2012, and in Saudi Arabia it went from 0.9 to 1.5 percent. However, Oman maintained the same level of arable land and agricultural activities.

Farming in Qatar

As the hydro-climatic limitations of growing certain crops were starting to affect policy decisions, Qatar was embarking on an ambitious intensive program of food self-sufficiency, one that is based on innovative science and technology-based farming methods. Its perspective is one of concern about geopolitical and socioeconomic risks such as regional military confrontations that would affect trade routes, and interruption of supply from major food-exporting countries. Qatar, therefore, decided to follow a mostly homegrown solution to its pursuit of food security.

The uphill battle that Qatar chose to pursue would, if won, enhance its national security and inform impoverished communities located in arid areas of the world on how to increase food production.[4] The challenges that face this effort are immense. For example, depleting aquifers and declining yields, and the higher farming costs are "forcing several farmers to quit farming" (Kanady, 2013). The national plan for food security as envisaged by the Qatar National Food Security Program (QNFSP) wants to boost the country's capacity for food production, preserve the nutritional quality of the produce, cushion consumers from food price volatility, and to make sure that farming is economically viable. It plans to have substantial private-sector participation, desalination plants intended for the farming sector, and programs that would train traditional farmers on modern farming methods. Its original goal was 70 percent food self-sufficiency by the year 2023, a goal that was recently downscaled to 40–60 percent with the timeline being more open-ended (Kanady, 2013; Peninsula, 2011a). This change of plans in terms of deliverables acknowledges the scale of challenges facing the QNFSP, which include capacity building among the current generation of Qataris who view the farm as place to escape to on the weekend, and whose career paths rarely include farming (Woertz, 2013).

---

[4] Despite the capital-intensive approach of Qatar's agricultural endeavors, some components of this effort are likely to be economically feasible for poorer countries.

The biggest expansion, however, was the development of extensive irrigation programs. From 1985 to 1995, the area under irrigation in Qatar went from 5,000 to 13,000 hectares, and in Kuwait from 2,000 to 5,000 hectares. In 1965, each of the Gulf states had 3,000 hectares or less under irrigation, with the exception of Saudi Arabia, where 353,000 hectares were already being irrigated (Bazza, 2005). In the span of three decades (1965–1995), the area under irrigation in Saudi Arabia increased by more than three-fold to reach 1,620,000 hectares, and went up at an even faster pace in the other Gulf states (FAOstat, n.d.). In the same time period, the area quadrupled in Bahrain to reach 4,000 hectares in 1995, tripled in Oman reaching 62,000 hectares, and doubled in the UAE to a total of 68,000 hectares . While all GCC countries experienced a massive expansion of lands under irrigation, the scale in Saudi Arabia was considerably larger because of the sheer size of the kingdom.

As this enlargement was occurring, the agricultural sector went from being an important source of employment to a marginal one, all in the span of about four decades. In 1978, the economy of Oman was very traditional and labor intensive. Farming employed about 70 percent of the active labor force but contributed only 2.5 percent to the GNP (Halliday, 1978). In 1996, it employed 40 percent of the labor force and yet accounted for less than 2 percent of the GDP. Currently, despite significant public investments, agriculture contributes less than 1 percent of the GDP of the Gulf countries, with the exception for Oman and Saudi Arabia. Of these two countries, the agricultural sector is strongest in Saudi Arabia yet it contributes a meager 2.7 percent to its GDP, and employs 4.7 percent of its work force (Table 3.2). Since the discovery of oil, the relative economic size of this sector has been shrinking rapidly. For example, in the span of one decade in the UAE, employment in agriculture dropped from 7.9 percent in 2000 to 4.9 percent in 2005 (data.UN.org) to 4.2 percent in 2010. Furthermore, the majority of those who do the actual farming throughout the Gulf tend to be foreigners (FAO, 2008), except in Saudi Arabia and Oman, where these workers play a somewhat smaller role. These realities raise questions about the social and economic value of farming for the Gulf states.

To summarize, until the early years of the twentieth century, the agricultural sector provided a certain level of food self-sufficiency for the sparse and impoverished populations of the Arab Gulf. Despite its expansion and modernization, this sector has been providing a diminishing percentage of the countries' food needs, mostly because of changing demographics, dietary habits, and disappearing groundwater. This has necessitated the expansion of the agricultural sector beyond the region; the broader foodshed framework helps to focus the discussion on food system security (Peters *et al.*, 2009). It is worth noting that there is resistance among some Gulf natives to the idea of globalizing the food supply system of their countries because it forces them into a position of vulnerable dependency.

Table 3.2 *Agricultural realities in the Gulf states*

	Arable land as a percentage of land area				Country total area** (thousands of ha)	Agriculture's contribution to GDP (2010; percent)	Employment in agriculture (as percent of total employment)***
	1982	1992	2002	2011			
Bahrain	2.9	2.8	2.8*	1.8	76	0.5	n.d.
Kuwait	0.1	0.2	0.7	0.6	1,782	0.3	2.5
Oman	0.1	0.1	0.1	0.1	30,950	1.4	n.d.
Qatar	0.5	1	1.0	1.2	1,159	0.1	2.7
Saudi Arabia	0.9	1.7	1.7	1.4	214,969	2.7	4.7
UAE	0.2	0.5	0.9	0.6	8,360	0.9	4.2

Source: FAO, Bloomberg as quoted by Alpen Capital (2011); World Bank Data,[5] accessed on March 5, 2012; see http://data.worldbank.org/indicator/AG.LND.ARBL.ZS?page=5.
* From 2003 and onwards, Bahrain's arable land drops to 1.4 percent of the land area.
** FAO, AQUASTAT.
*** For various years in mid 2000s (World Bank Data).

Food security has been on the minds of decision makers in different Gulf states who have been contemplating the idea of extending their agricultural activities to fellow Arab countries for decades. More recently, in early 2008, Jacques Diouf, the Director of the United Nations Food and Agriculture Organization (FAO) said to Arab ministers in Cairo that because food prices have risen and are likely to stay high, the world may face a food-availability problem even if countries have the resources to purchase it. He then concluded that these countries ought to "consider the possibility of investing in some of their sister countries ... to ensure the security of their supply ... and in the same vein to help the agricultural development of these countries, which would be a win–win situation" (Saleh, 2008). The countries that Diouf had in mind were African countries, especially Sudan. Furthermore, Shamshad Akhtar (2011), the World Bank's vice president for the Middle East and North Africa, said that the "limited water resources in the Arab world limits potential for domestic food production" and called for a policy approach that is based on managing water demand. This recommendation was based on the fact that the majority of the region's water originates outside its borders, and that climate change is forecast to change precipitation patterns and cut per capita water availability in half by 2050.

Natural environmental conditions set the limits of what is agriculturally feasible. Human ingenuity and technological innovations have been able to expand

[5] The World Bank uses the FAO's definition of arable land to include land under temporary crops (double-cropped areas are counted once), temporary meadows for mowing or for pasture, land under market or kitchen gardens, and land temporarily fallow. Land abandoned as a result of shifting cultivation is excluded (World Bank Data, n.d.).

these limits, but at substantial economic, and sometimes ecological, costs. Western Asia, the name of choice used by the United Nations for the "Middle East," experiences severe agricultural constraints where poor soil and aridity limit crop production (Fischer *et al.*, 2002). These extensive constraints necessitate substantial irrigation; the agricultural sector consumes 80 percent or more of all the freshwater used in the Gulf states. The anomaly, however, is that water's contribution to the GDP and to employment is truly minimal. As the Gulf governments started investing in and expanding their modern agricultural sector, signs of an ecosystem degradation began to appear. As intensive commercial agriculture was starting to take root in Oman in the late 1970s, there were signs that the country's scarce water resources were being depleted rapidly and its aquifers may have been damaged "irrevocably" (Halliday, 1978). Kuwait and Saudi Arabia had similar experiences. Furthermore, the UN (FAO, 2008) reports that increasing soil salinization, a growing phenomenon in the Gulf states, is removing land from farming and thus shrinking the area under cultivation (Table 3.2). This salinization process, sometimes a byproduct of irrigation in hot and dry climates, is having a strong but variable effect on the soils of the Gulf states. At the worst end of the spectrum, 85 percent of Kuwait's farmland is salinized (FAO, 1997). In Bahrain, the area of agricultural land dropped from 6,460 to 4,100 hectares between 1956 and 1977, a decline that was attributed to urbanization, water logging, and salinization (FAO, 1997).

Such soil degradation is due to poor management of desert irrigation, which includes the excessive use of water and the relative shortage or absence of drainage. Recently, extensive land reclamation programs in most Gulf states have succeeded in increasing the net area of farmland. This, however, is putting pressure on aquifers, which are being overdrawn to the point that soil salinization in recently developed lands is being attributed to the poor quality of irrigation water that is caused by seawater intrusion into subterranean waters. A new United Nations report (WWAP, 2012), which ranked countries with the highest rate of groundwater abstraction, found Saudi Arabia to be eighth in the world. Furthermore, the climatic conditions in which food is being produced require the crop to be dependent on irrigation. The cultivation of alfalfa (*berseem* in Arabic) as fodder was popular because farmers were charged very little for irrigation and because its yield is high (74.5 tons/ha) when compared with that of vegetables like tomato (11.7 tons/ha) and fruit trees like dates (7.5 tons/ha). Also, alfalfa generates multiple crops because it can be grown throughout the year and tolerates water with higher salinity levels (FAO, 1997). Therefore, from the point of view of farm owners, alfalfa is a lucrative crop to grow. By the mid 1990s, fodder crops like alfalfa were grown in each of the GCC countries where they occupied anywhere from 12 percent of the cropped lands in Saudi Arabia to 32 percent in Qatar (FAO, 1997). Alfalfa, however, is a far thirstier crop than wheat as it consumes five times more water, and even greater quantities during the summer (Woertz, 2013). Given this hydrological and climatic context, the remaining sections of this chapter will analyze the local and global ramifications of farming abroad for the Gulf states.

3.4 Food security through farming abroad

The primary goal of the expansion and modernization of the farming sector is to provide the conditions necessary for increasing food output to meet the food needs of the local population. Food security is a multi-dimensional concept that includes environmental, economic, political, policy, and cultural variables (Sen, 1981; IMF, 2012b). People who are unable to meet their daily nutritional food needs are more likely to experience failing health, to require medical attention, and to be unable to work to their full potential, which would result in a drag on the national economy. The 2007–2008 global surge of food prices around the world triggered riots in over 60 countries that stretched from Bangladesh to Burkina Faso (CSIS, 2010; Donald *et al.*, 2010), which is consistent with historical experiences from the French Revolution to the Arab Spring (Sternberg, 2013).

Food security, according to the FAO (2002), is "when all people, at all times, have physical, social, and economic access to sufficient, safe and nutritious food that meets their dietary needs and food preferences for an active and healthy life." Therefore, it is about the availability, access, and stability of food supplies, with the latter criteria having been adopted during the 2009 World Summit on Food Security. In the late 1970s and 1980s, the Gulf states experienced a significant improvement in their economic conditions. This unleashed an even bigger inflow of foreigners, diminishing the ability of the agricultural sector to meet the people's food needs with domestic sources. Governments took steps to adapt to the unfolding environmental and demographic pressures in order to reduce their vulnerability to disruptions in water and food supplies.

Many Gulf states' governments decided to enhance their security by growing food on "their own" farms abroad (Table 3.3). The UAE's investments in agriculture abroad increased 45 percent between 2006 and 2008 (Lowe, 2011). Gulf countries like Saudi Arabia, Kuwait, and the UAE have decided to use their financial wealth to enhance their food security. Their aim was to invest in countries that they have good diplomatic and economic relations with, are geographically close in order to reduce transportation costs, and are endowed with agro-climatic conditions suitable for efficient, large-scale farming activities.

The diversity of investment locations acts as insurance against the risk of cascading effects of supply disruptions. For a country to buy or lease farmland outside its national borders is nothing new. However, as global food prices began to rise in 2006, the once anemic investment in farmland became hyperactive as its pace and magnitude accelerated significantly. Land Matrix Partnership, a European

Table 3.3 *GCC countries' overseas land investments*

GCC Investors	Host countries	Stated purposes for projects	Scale of deals
Saudi Arabia	Ethiopia, Sudan, Senegal, South Sudan, Russia, Philippines, Argentina, Egypt, Mali, Mauritania, Nigeria, Niger (suspended by host in 2009), Pakistan, Zambia	Direct export of maize, soybean, fodder, rice, palm oil, prawns, bananas, pineapple, vegetables, wheat, poultry	Of these deals, 16 cover 1,713,357 ha. Five of these are in Ethiopia
UAE	Sudan, Algeria, Morocco, Egypt, Ghana, Indonesia, Namibia, Pakistan, Romania, Spain, Sudan, Tanzania	Direct export of potatoes, olives, dairy, olive oil, citrus, fodder, maize, palm oil, rice, sugar cane, dates, alfalfa, cereals, cotton, sunflowers, peanuts, sorghum	Of these deals, five cover 1,882,739 ha
Qatar	Cambodia, Sudan, Turkey, Brazil, Vietnam, Pakistan, India, Ghana, Indonesia, The Philippines, Australia	Direct export of sheep, wheat, cereal, rice, barley	Of these deals, four cover 642,630 ha
Kuwait	Cambodia, Laos, The Philippines	Direct export of rice and maize	
Bahrain	The Philippines	Direct export of bananas and rice	
Oman	The Philippines	Direct export of rice	

Source: Report on "GCC states" Overseas Land Investments, 2012 as quoted in Alpen Capital (2013), see http://www.alpencapital.com/downloads/GCC_Food_Industry_Report_May_2013.pdf.

research group, reported that in developing countries "as many as 227 million hectares of land – an area the size of Western Europe – has been sold or leased since 2001, mostly to international investors." It also notes that most of these acquisitions took place from late 2009 to 2011 (Oxfam, 2011, p.2). Based on data for 2006–2009, the International Food Policy Research Institute (IFPRI) ranked countries by the area of land leased or owned abroad. The top three were China, South Korea, and the UAE (Lowe, 2011).

The global land rush became so pervasive that countries, Arab and non-Arab, large and small, and even individuals like Phil Heilberg, a former Wall Street trader, have been acquiring lands in faraway countries (Table 3.4). In 2009, Heilberg leased one million acres of farmland in the war-ravaged savanna of southern Sudan, a tract nearly the size of the American state of Delaware; this

Table 3.4 *Land deals abroad as a percentage of agricultural area*

DR Congo	48.8
Mozambique	21.1
Uganda	14.6
Zambia	8.8
Ethiopia	8.2
Madagascar	6.7
Malawi	6.2
Mali	6.1
Senegal	5.9
Tanzania	5
Sudan	2.3
Nigeria	1
Ghana	0.6

Source: FAOstat as quoted by BBC (2012).

made him "one of the largest private landholders in Africa" (Funk, 2010). He was hoping to acquire another million acres of very fertile land located in the south-eastern part of the country, and irrigated by a tributary of the Nile River making it drought-proof (Funk, 2010). In cases like this, foreign governments that own large tracts of land or hold long-term leases to them could become "virtual riparians" on a transboundary river like the Nile, and this would significantly complicate riparian relations along transnational river systems. Between 2005 and 2010, Heilongjiang Beidahuang Nongken Group, China's largest agricultural company, had invested more than 250 million yuan ($38 million) in overseas projects and was planning to expand its investment further (Yan and Chang 2011). In 2011, it was working on acquiring 200,000 hectares of farmland in Russia, the Philippines, Brazil, Argentina, Australia, Zimbabwe, and Venezuela. The group follows different business models depending on the country that they are in. In Venezuela and Zimbabwe, it provides machinery and laborers, and in return it takes about 20 percent of the harvest. While it acquired farmlands in Australia, it leased them in Brazil and Argentina (Yan and Chang, 2011).

Countries that are poor or have governance problems could benefit from a close relationship with wealthier and more powerful countries, especially if they share common cultural traits. Foreign direct investment, even in countries with high levels of corruption, could stimulate the national economy by creating jobs, often higher-paying ones. The Pakistani Prime Minister Yousuf Raza Gilani offered hundreds of thousands of acres of agricultural land to the Saudis in return for oil (Agence France-Presse, 2008b), and the Senegalese government offered them 400,000 hectares (Pearce, 2012). The Sudanese government decided to set aside one-fifth of the country's cultivated land for the purpose of them being used by Arab countries (Economist, 2009).

Using satellite imagery data from 1995–1996, a comprehensive survey (Fischer *et al.*, 2002) of global agriculture found that 80 percent of the world's reserve agricultural land is located in Africa and South America, and estimated the total cultivable land in Africa to be 807 million hectares, of which a mere 197 million hectares are under cultivation. This leaves the impression that much of these (sub) continents are virtually empty. However, for rural residents where land concessions have been made to foreign investors, unused lands and national conservation areas are appropriate fields for livelihood pursuits (Li, 2011). Sometimes, "unused" lands are farmlands that have been left fallow to give them time to regenerate their nutrients. Also, land use by nomads is occasional or seasonal where their herds use the pastureland for grazing a few weeks a year.

International investments in foreign lands were accelerated after the 2008 global surge in food prices, which in turn led some 25 countries, including major grain exporters like Vietnam, Argentina, and India, to ban or impose restrictions on cereal exports (Mousseau, 2010). The decision to look outside the Gulf region to bolster food security is primarily related to existing hydro-climatic conditions, and the rapid depletion of fossil aquifers. For instance, the Saudi government's grain-buying agency, the Grain Silos and Flour Mills Organization (GSFMO) has decided to reduce domestic wheat purchases by 12.5 percent annually until the wheat production program comes to an end by 2016. This, for Saudi Arabia, is a dramatic water-saving move that extends the life of its aquifers, which they now view as strategic reserves. This boosts the country's resilience as the aquifers could be vital in times of hydrological emergencies. By 2015, Saudi Arabia is expected to import around two and half million tons of wheat for human consumption. The livestock industry, which currently depends on barley for most of its animal grain feed, is considering replacing some of that with wheat, a move that would boost wheat imports (OBG, 2013). As the GSFMO continues to purchase wheat directly from farmers and from the international market, officials like the minister of agriculture, Fahd Balghunaim, have been encouraging "private businesses to buy or lease land overseas with the aim of producing crops for domestic consumption." For its part, the Saudi government has pledged to work with other governments to ensure investment security in other markets. The minister described private sectors' international investment in the agricultural industry as being essential for ensuring the country's food security (OBG, 2013).

The secondary, yet important, driver is the historical experience of the Gulf states. For example, in 1973, the Arab oil-exporting countries, mostly concentrated in the Gulf region, organized an oil embargo aimed at getting Western governments to force Israel to return Arab lands that it had militarily occupied during the

1967 (i.e. Six-Day) War. The national security advisor to the American president, Henry Kissinger, pressed the Arab states to lift the embargo, and signaled that the United States would retaliate. A few months later, he suggested that the United States was willing to implement a "counter embargo" and in effect use "food as a political weapon" by halting food sales to the Arab Gulf states (Rothschild, 1976, p. 300; Woertz, 2013). Around that period of time, the US Secretary of Agriculture, Earl Butz, stated that "hungry men listen only to those who have a piece of bread. Food is a weapon in the US negotiating kit" (Patel, 2007, p. 91; see also Paarlberg, 1985). In December 1980, John Block, Ronald Reagan's Secretary of Agriculture, said "I believe food is the greatest weapon we have for keeping peace in the world" (Shepherd, 1985). Consistent with this policy and as a reaction to the Soviet Union's invasion of Afghanistan, the United States imposed a grain embargo (February 1980–April 1981) on its Cold War adversary. Another concern of Saudi planners is that Western countries could block food imports into the country in order to influence its oil policy (Jones, 2010). These historical experiences focused Gulf countries' attention onto the issue of food security, which influenced Saudi Arabia's decision to launch a program of wheat self-reliance, and nudged the Gulf states to consider farming abroad.

The Oxford Business Group (OBG, 2013) reports that the Saudi private sector and the government have made land investments in countries that include Brazil, Canada, Ukraine,[6] Ethiopia, and Sudan. Combined, their investment is worth more than $11 billion in agricultural ventures (OBG, 2013). The United Farmers Holding Company (UFHC), a Saudi consortium made up of Saudi Agricultural and Livestock Investment Company, Saudi Grains and Fodder Holding, and the kingdom's dairy giant, the Almarai Company, have taken preliminary steps to purchase the Irish-based agribusiness Continental Farmers Group (CFG). The transaction would give UFHC control of 2,700 hectares in Poland and 33,000 hectares in Ukraine, land that would be used to grow wheat. As Saudi Arabia increases its reliance on food grown outside of its borders, the agribusinesses will play a larger role in enhancing its food security.

The Economist magazine (2008b) argued that Saudis are buying farmland abroad in order to have greater confidence in the "security of their food supplies". They are injecting capital, and can "offer cheap fertiliser, which it can produce by using subsidised gas." The concluding sentence states that Saudi investors "may be resented for buying up primary commodities from poor countries, while monopolising and limiting the output of their own special one: oil." While Saudi Arabia is

---

[6] The only Arab country to have agricultural investments in Ukraine is Saudi Arabia. Collectively, the Public Investment Fund (PIF) of Saudi Arabia, Saudi Al Rajhi Group, and Almarai Co own 33,000 hectares (The Oakland Institute, 2014). There is no indication that the turmoil that has rocked Ukraine since the fall of 2013 is affecting the government's policy towards foreign land investments. This, of course, may change with time.

the largest producer of the Organization of Petroleum Exporting Countries (OPEC), the organization produces 40 percent of the world's oil output yet exports 60 percent of total petroleum products traded globally. Hence, neither the country nor the organization's actions were judged as being able to, and "do, influence international oil prices" (EIA, n.d. a). Furthermore, while OPEC regulates its output it is not in the business of banning exports or antagonizing consumers. For example, when in 2008 the Iranian president Mahmoud Ahmadinejad urged the OPEC to stop pricing oil trades in US dollars, its members dismissed it as a stunt. A year earlier, as the geopolitical rhetoric about Iran's nuclear program was heating up, the same leader of this major oil- and gas-producing country and member of OPEC said that if his country's nuclear infrastructure were attacked by the United States, it "would never like to use oil as a weapon . . . there are other means at our disposal to respond" (Reuters, 2007). As the economies of the Gulf states have become deeply integrated into the global economic order, their leaders realize that economic growth is fueled by political stability and cooperation, not by belligerence and confrontation.

3.5 Colonialists or investors?

Critics label foreign investments in agriculture as neocolonialist and reminiscent of the "banana republics," those weak and authoritarian governments "whose econ-omies are dominated by foreign-owned fruit plantations" (Economist, 2009). Currently, the alleged "closet" neocolonialists, the Gulf states, were themselves colonies some four decades earlier. Except for Saudi Arabia, the Gulf states have very small population sizes and tiny armies (Figure 3.2). Hence, their security is ensured by a close alliance with the United States. Their political stability seems to be continuously challenged by various geopolitical tremors that rock the region, and their economies are dependent on the fluctuating international price of a single export commodity (Table 3.5). Such countries are neither interested in, nor want to be seen as neocolonialists. Foreign land investments provide host countries with much-needed agricultural capital, create jobs, and boost food supply because some of the produce gets sold locally. For example, eggs produced by China-owned enterprises are sold in Zambia, and Sudan forbids agricultural investors from exporting 30 percent of their produce and requires them to build infrastructure projects. Some 30 percent of the rice produced by Saudi Arabian firms on Senegalese land is sold in local markets (Pearce, 2012). Qatar is building a seaport in Kenya in exchange for access to farmland, and some of its farming projects are in collaboration with local governments such as the Qatar–Sudan investment joint venture. An agricultural official in Sudan said that Arab governments' investments in his country would go from $700 million in 2007 to $7.5 billion in 2010, or from

Table 3.5 *Impacts of price volatility*

Channel	Who/what is affected?	Examples
Poverty traps	Consumers and farmers	Temporary coping mechanisms such as distress asset sales or reduced intake of nutritious foods leading to permanent effects
Reduced private farm-level investment	Farmers	Lower fertilizer use leading to lower productivity
Macroeconomic impacts	Volatile food prices reduce the ability of prices to function as signals that guide resource allocation	Investment not directed to optimal sectors of the economy, reducing economic growth
Political processes	Democratic institutions; long-term economic growth	Food riots that damage investment climate; subsidies that prevent investment in public goods

Source: FAO (2011b).

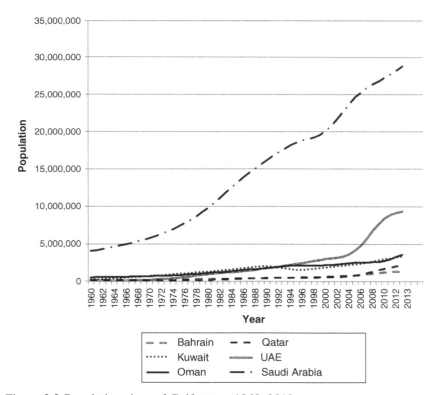

Figure 3.2 Population sizes of Gulf states, 1960–2013

3 percent to around 50 percent of all the investments in the country (Economist, 2009).

In the mid 1970s, the Arab countries wanted to convert Sudan's massive yet idle or under-utilized farmlands into a "bread basket of the Arab World." The war and instability in Sudan, especially in the south, undermined what seemed like a rational Arab solution to an Arab concern. Sudan today does not use its share of the Nile water under the 1959 water allocation agreement, hence the government "plans to pay more attention to agriculture as a focal point of its national economic strategy. This in turn will mean the need for more waters than Sudan is currently using. Sudan has large irrigable lands that have hitherto not been developed, and it has recently revived the four-decade-old slogan of Sudan being the breadbasket of the Arab world." (Salman, 2011).

Therefore, the idea that some arid Arab countries would outsource their food-producing capacity is well known and dates back to a much earlier era. Despite some narrow feelings of nationalism, internal differences, and some transnational disagreements, Arabs nevertheless share certain feelings of brotherhood, if not as Muslims, then certainly as fellow Arabs. Agricultural interdependence is in effect a venue through which Arabs can more conveniently display their brotherhood without getting marred by the heavy mud of politically charged ideas of Arab nationalism and Arab unity (Jones, 2010). In addition to the intangibles of brotherhood, the GCC countries decided that farms abroad needed to be closer to the home turf. This explains not only the focus on Arab and Muslim Sudan, and Muslim Pakistan and Indonesia, but also on the non-Muslim Philippines, Vietnam, and Cambodia (Table 3.3). In other words, the spatial distribution of these lands shows that countries' locations and hydro-climatic attributes have played a deciding role in where to acquire land. This shows that decision makers in the Gulf states are pragmatists, not cultural ideologs; and globally engaged capitalist investors, not neocolonialists.

Historically, colonial powers were accused of using territories that were under their control to grow cash crops that were exported to the homeland without much benefit to local populations. The concern is that current investors' schemes may have similar exploitative effects on host countries who are also quite poor. The distinguishing criteria of current land acquisitions are their larger sizes, the strong influence that host and investing governments play, the general lack of transparency, and the business-friendly environment in which they are located (Niasse and Taylor, 2010). However, investors' purchases or leases of farmlands are estimated to be worth $20 to $30 billion, ten times greater than the World Bank's emergency package for agriculture (The International Food Policy Institute (IFPRI) as quoted by the Economist (2009)).

Saudi Arabia's initiative to farm abroad was launched by King Abdullah himself; its main aim is to grow different cereals such as rice and wheat for the

Saudi market. The media was critical of these investments. The Economist, a British cum global magazine, wrote that the Saudi government and the World Food Organization are each spending about the same amount, $100 million, the former on producing food in Ethiopia, the latter on food aid to the same country (Economist, 2009). It added that "the Saudi Program is an example of a powerful but contentious trend sweeping the poor world: countries that export capital but import food are outsourcing farm production to countries that need capital but have land to spare."

Foreign agricultural investments of the Gulf Arab states, especially those of Saudi Arabia, appear to receive significant attention and scrutiny from the world's media, much more so than similar investments by American investment banks, and Chinese or Israeli firms These Arab countries are investing their wealth in agricultural projects, much in the same way as other non-Arab countries are doing. Furthermore, extractive industries such as oil, uranium, and gold have always attracted foreign direct investments, often in troubled lands run by corrupt regimes. However, unlike these depleting activities, farming is a renewable activity if production processes are environmentally, socially, and economically sustainable. Critics of farming abroad argue that foreigners who may have leased the land for a period of time do not have an incentive to sustainably manage it, and their use of the land could degrade it, therefore making it akin to an extractive industry. Some argue that investors in foreign lands need to respect customary rights, share benefits among locals by creating jobs for them, increasing transparency in how contracts are negotiated, and respecting national trade policies, which forbid exporting food if the host country is experiencing a famine. The impacts of these investments on the agricultural sector of host countries are an important indicator of whether these investments are beneficial to local areas or not.

Neocolonialism is related to the power differential between an investing country and the one hosting it, and the effects of that on the local population. Were an authoritarian government to submit to pressure and consult its native population, fear, intimidation, and corruption would slant the results in favor of the central government. Furthermore, peasants in deeply impoverished countries like Ethiopia, Sudan, and Madagascar have low levels of human development, and are therefore very disadvantaged with respect to investors or national governments. Consequently, they could be manipulated, their rights squandered, and some might be even displaced. Similarly, corrupt officials undermine the national interest and the legitimacy of the government. Widespread corruption, as is the case in many countries (Transparency International, 2013), is an indicator of failed governance, and speaks to the need to bolster political and legal institutions, and to add transparency in decision-making processes, which would help protect the rights of citizens.

Table 3.6 *Advantages and disadvantages of land transactions*

Pros	Cons
Modernization of long-neglected agriculture through investments	Secret deals (Kuwait–Cambodia rice deal)
Agricultural research and development in target countries (e.g. Chinese research stations in Africa)	Blackmailing target governments (e.g. China threatened to pull out of Zambia if opposition leader became president. He was against biofuel arrangement for the Chinese).
Genetically modified seeds suitable for local conditions	Dispossessing local peasants
New or improved infrastructure (e.g. roads, ports, water wells, irrigation systems, and so on)	Degrading water and soil
New markets	Displacing local population
Better jobs	
Services (clinics, schools)	

Based on the Economist (2009).

To characterize the globalization of farming as "neocolonialist" and as land "grabbing" is to sensationalize the debate and polarize the stakeholders, which does not help in clarifying the costs and benefits in the short, medium, and long terms.[7] (Table 3.6) Such emotive labels exaggerate the misdeeds of sudden land rush by corporations and state-sponsored investors. Significant power differentials between the parties involved and the varying political cultures make for many unique circumstances, which makes it harder to judge many of these international transactions. In the interest of charting a common pathway foreward, European advocacy groups (Saracini, 2011, p. 5) posit that when one or more of the following factors are present, a land acquisition would be considered a land grab if:

(1) there is violation of human rights, and particularly the equal rights of women;
(2) there is no involvement of free, prior, and informed consent of the affected land-users;
(3) it is not based on a thorough assessment of the consequences, or are in disregard of social, economic, and environmental impacts, including the way they are gendered;

---

[7] These impacts could be social, environmental, and sometimes political, and could affect a certain region, ethnic, or religious group, but not another. In short, sweeping statements that paint all land investments as land grabbing are inaccurate; contents and parameters of transactions are often not made public, vary from one country to another, and take place in differing ecological, political, social, and economic environments. Finally, the benefits of foreign land investments are hardly considered by those who research the topic.

(4) it is not based on transparent contracts that specify clear and binding commitments about activities, employment, and benefits sharing; and

(5) it is not based on effective democratic planning, independent oversight, and meaningful participation.

These are reasonable expectations to have of host countries and of foreign investors. Hardly anyone can argue against the need for transparent and participatory process in how land transactions are carried out. However, it is difficult to imagine how a foreign investor can ensure equal rights of women under different cultural and political conditions. The criteria offer admirable normative goals and values that should be taken seriously, and be used to shame investors publicly into abiding by an ethical framework that is based on transparency and protection of people's rights. While it is difficult to stop the forces of the global marketplace, non-governmental organizations ought to do what they can to channel agricultural investments towards a more inclusive and transparent process in the host country. Realistically, however, because some of these criteria may be difficult to implement at the present time, investors should therefore treat them as higher goalposts to strive towards. The investors' primary goal is to meet their business objectives; these tend to be financial gain for shareholders or food security for the investing country, or both; however, the goals change over time. The views of some Western commentators on land acquisitions are rather idealistic (De Schutter, 2011). Their starting point is justice and equity for the people of the host country. For example, some research institutes have recommended that investors should be banned from exporting food products if the host country is experiencing an acute food-deficiency. Principles like these stand on ethical and humane bases. However, implementing them is sometimes challenging because many host governments have high levels of corruption and repression where the people are denied fundamental freedoms.[8]

Transparency International's Corruption Perceptions Index measures the perceived levels of public sector corruption in 177 countries worldwide, and uses a ranking scale from 0 (highly corrupt) to 100 (very clean). Its 2013 index reveals that more than two-thirds of the countries surveyed score less than 50, and that corruption in those that have been the target of land grabbing varies widely. While most countries that have been targeted for land investments have very low transparency (scores below 40) some, like Turkey and Poland, are transparent and harbor lively, free media (Table 3.7). In the same way, the UAE's and Qatar's business operations tend to be transparent, while those in some other Gulf states are less so. It should be noted here that media outlets in the Gulf states are mouthpieces of the ruling families and are not likely to investigate and publicize corrupt transactions.

---

[8] Some investing countries fall into that category as well.

Table 3.7 *Corruption Perceptions Index in select countries*

Country	Index	Country	Index
Sudan	11	Poland	60
Ethiopia	33	Canada	81
Egypt	32	Turkey	53
Ukraine	25	Argentina	34
Vietnam	31	Brazil	42
Pakistan	28	Saudi Arabia	46
Zambia	38	UAE	69

(Transparency International, 2013).

Recent leaks of US Embassy documents from early 2011 reveal that the government of Ethiopia sold large parcels of land to the former president of Nigeria, Olusegun Obasansjo, the sitting president of Djibouti, Ismael Omar Guelleh, and to the prime minister of Egypt, Ahmed Nazif. Unlike the first two parcels, which were purchased for personal use, the latter was purchased on behalf of the Egyptian government. It leased 49,400 acres of land in the Afar region to grow cereals for export to Egypt. The Egyptian, Djibouti, and Saudi tenants were quietly exempted from the official ban on the export of cereals which the Ethiopian government had put in place in the aftermath of the spike in food prices. The official cable confirmed that other investors "have not been allowed to export cereal grains"(Afrol News, 2012).

The discourse on land acquisitions is dominated by a narrative that is somewhat charged. For example, De Schutter's 2011 commentary reveals his stance in the first few lines of his essay, where he states that "the real concern behind the development of large-scale investments in farmland is that giving land away to investors, having better access to capital to 'develop', implies huge opportunity costs as it will result in a type of farming that will have much less poverty-reducing impacts . . ." (De Schutter, 2011, p. 249). Then the author defines "land grabbing" as the "acquisition or long-term lease of land by investors" and hints that its continued use is wrong and unethical. Authors with such a perspective routinely use this phrase throughout their work as a given, and not as a finding or conclusion of their research. In other words, it would be consistent with accepted methodologies in social science if researchers "concluded" that, given some realistic criteria, certain types of foreign agricultural investments amount to land grabbing. Sweeping, sensational descriptors like "land grabbing" can be distracting and polarizing, at a time when an inclusive debate on the issue would inform the various stakeholders and better serve their interests..

The problem is, however, that some investors have paid paltry sums for acquiring farmland, and that the majority of transactions are signed quietly, and away from public scrutiny. This fuels suspicion and hearsay. In a few host countries, when locals protested land deals that were arrived at with foreign investors, government forces dealt with them violently. Furthermore, some of the lands that were sold or leased are occasionally used by nomads, herders, and gatherers who don't have a formal, legal title to the land. Early in 2008, when food prices were rising, Bahrain announced plans to purchase 300 million square meters of agricultural land in eastern parts of Saudi Arabia. The land was to be rented out to Bahraini farmers who would produce food for the people of Bahrain (Gulf Daily News, 2008). Bahrain's Municipalities and Agriculture Ministry was expected to provide water, electricity, and seeds. This was expected to create one thousand jobs for Bahrainis and, most importantly, it was conceived of as a step towards achieving food security for their country. The two monarchies have warm diplomatic relations, and are geographically adjacent; the island kingdom became physically joined with Saudi Arabia in 1986 by a 25-kilometer-long bridge. It turned out that 2008 was a pivotal year for agriculture in Saudi Arabia. As mentioned above in this chapter, the kingdom announced a restructuring program, which included a gradual phasing out of wheat farming by 2016.

Although Bahrain's plan fizzled, the very idea demonstrates the perception among the people of the Gulf about Saudi Arabia being the region's "garden," and illustrates countries' natural preference to farmlands in nearby and friendly countries. It also demonstrates an outlook that is pragmatic, functional, and non-exploitative. Similarly, Qatar and Bahrain, the smallest in area and population among the Gulf states, have been exploring opportunities to lease or purchase agricultural lands in various countries around the world including Australia, India, Pakistan, the Philippines, Egypt, and Sudan. (GRAIN, 2008). The efforts of the Gulf states could not have neocolonial motives, especially Bahrain, the oil-free, poorer, and politically troubled Gulf country. Furthermore, the population of Qatar is 300,000 natives, and over a million guest workers. Ascribing neocolonial motives to the Gulf states is a stretch; even if they are using their financial wealth to extend their foodshed beyond the homeland, and consequently ensure their food and water security.

Some major international agricultural investors like China have adopted an approach that creates jobs for their own nationals in the host country. China has a "going out policy" which encourages greater levels of investments in other countries, the result being that a larger number of Chinese workers are deployed overseas. China's Ministry of Commerce figures show that there were 812,000 workers abroad at the end of 2011, [9] which is twice the level of 2002, and foreign investments excluding the financial sector were at $60 billion (Forsythe, 2012).

---

[9] China evacuated 35,860 of its citizens from Libya during the 2011 revolutionary war there (Forsythe, 2012).

The concern is that investing states are not likely to create many employment opportunities for native people, a point that was echoed in a research paper on the effects of land grabs on local labor. It argues that, in light of the vast unemployment levels in the global south, and "unless vast numbers of jobs are created, or a global basic income grant is devised to redistribute the wealth generated in highly productive but labor-displacing ventures, any program that robs rural people of their foothold on the land must be firmly rejected" (Li, 2011, p. 281).

According to Lorenzo Cotula of the International Institute for Environment and Development (IIED), European and North American corporations are involved in many more land deals abroad than China or the Middle East. At the local level, Cotula said that villagers have different views on the impacts of land deals; some would benefit from the land deals while others would be harmed. Therefore, there is no framework or template for foreign agricultural investments that could be implemented in every situation. His assessment is that their overall impact is likely to be negative (Berger, 2013). In 2012, the FAO of the United Nations proposed voluntary guidelines for responsible governance of natural resources. The document repeatedly stresses the need for governance process to be participatory and gender-sensitive, and that governments should "recognize the reality of the situation" of informal land tenure and prevent corruption and forced evictions from the land (FAO, 2012, p. 16). Local governments need to protect (in)formal land tenure, and to ensure that owners are compensated a fair market value for their property whether they elect to lease or sell. Such a framework would also mean that the landowner and would-be seller is fully aware of (un)employment and other ramifications of these transactions. Due to the prevalence of corruption and the significant power differential, including financial and educational deficits in the host countries, exploitation is likely to continue.

International investors sometimes have to deal with corrupt and/or autocratic foreign governments or transnational corporations. These experiences have typically been in poor and politically fragile countries, and have become part and parcel of humanity's modern economic history. Western oil companies have long invested in oil fields in troubled lands where they brought in capital, technology, and mercenaries and exported oil and profit. Well before agribusinesses started scanning the world for lucrative opportunities, oil and gas companies had identified hydrocarbon deposits in different parts of the world, including some in turbulent areas. The once-united Sudan was a poor yet oil-rich country that was unable to extract the riches of its lands without the investments of Western companies like Chevron (United States), Lundin (Sweden), Talisman (Canada), and others. After the discovery of extensive oil reserves in 1980 around Bentiu, these multinationals had "a non-negligible role" in re-igniting the conflict between south Sudan and the central government, and altered the balance of power in the country in favor of the repressive government (Schollaert and Van de Gaer, 2009). Compared to that, Arab agricultural investments abroad are rather benign.

The production and movement of oil sometimes have significant environmental and social impacts on local populations and ecosystems. The processes contaminate the soil and water, especially in areas where environmental regulations are lacking or unenforced. Other impacts arise from burn-off of excess natural gas, which releases methane, sulfur dioxide, and toxic compounds that pollute the air and water. The social consequences are equally substantial. The rapid flow of workers into an oil-producing region sometimes includes those from competing ethnic groups, or people who carry new diseases, bring social ills like prostitution, or, in some cases, are hardened prisoners. It was reported that of the 7,000 Chinese workers who were brought in to build the Port Sudan pipeline, about two thousands of them were "prisoners who were promised reduced sentences for their work" (Switzer, 2002; see also Wesselink and Weller, 2006).

In the oil-rich regions of Sudan, the central government cleared the land of civilians through a scorched earth policy, and the violent displacement of the people. The Sudanese army is said to have armed, funded, and deployed the Murahaleen bands of nomadic Arab tribes to protect the oil concessions. The United Nations Special Rapporteur to the Human Rights Commission reported, "the Murahaleen do not only target rebel camps or armed individuals, but also civilians, in a very intensive manner. Usually, food crops are destroyed, men are killed, and women and children are abducted." (UN, 2001). It was estimated that the government of Sudan had spent $300 million it had earned from the oil sector on weapons. Some argued that despite proclamations in support of ethical business practices, "human rights and development, the ugly truth is that Talisman is helping the government extract oil, and oil is paying for the war" (Economist, 2000).

In southern Sudan, armed insurgents killed three oil workers on the Chevron base near the town of Bentiu (Fisher, 1999). Regardless of the political and power dynamics, and said benefits associated with oil or land investments, if the locals are disaffected they are likely to rebel against investors; this would increase their cost of doing business or, in few cases, disrupt their operations completely. Similarly to corporate practices in Sudan, Nigeria was also affected by extractive industry's disregard for social and environmental protection that had adverse impacts on local communities. The Shell Petroleum Development Company of Nigeria[10] headed these operations in cohesion with the Nigerian National Petroleum Corporation (NNPC). Oil drilling operations by Shell have taken place in Nigeria since the 1950s. Boele *et al.* state that in the decades that followed, Shell essentially became

---

[10] The Shell Petroleum Development Company of Nigeria is Shell's subsidiary in that country.

the development agency within the Nigerian state, asserting itself as the dominant technological and economic resource entity. Although Shell did provide infrastructure developments such as roads, wells, and electricity lines to Nigerian citizens, the company failed to provide citizens with the most basic civil and social rights as well as a clean environment (Boele *et al.*, 2001, p. 130).

There has been significant evidence surrounding the source of devastation in Nigerian towns. Amunwa reports that Shell drove human rights abuse in Nigeria through employment contracts with armed rival militant gangs. According to an investigation by the oil industry watchdog Platform and other non-governmental organizations, Shell has been fueling this violence for more than a decade, causing fatalities as well as devastation of entire towns (Smith, 2011). This violence was promoted in order to ensure the protection of the Shell infrastructure. Essentially Shell distributed money to whichever gang controlled access to its infrastructure (Smith, 2011). This was specifically the case in the town of Rumuekpe, as it is the main channel for Shell's eastern operations, producing around 10 percent of Shell's daily production in the country (Smith, 2011). A gang member, Chukwu Azikwe, revealed that his group was given money, which it used to purchase food and ammunition, hence sustaining the war (Smith, 2011). Shell admits to these accusations by stating that they frequently invested money into the town of Rumuekpe, knowing that it was going towards the conflict (Amunwa, 2011). A Shell official confirmed these claims by ex-gang members who said that in 2006, "Shell awarded six types of contract in Rumuekpe. Thousands of dollars flowed from Shell to the armed gangs each month" (Smith, 2011). Shell contributed to human rights abuses in the country, and its oil spills and gas flaring in the Niger Delta had strong adverse effects on the natural environment (Amunwa, 2011).

The environmental impacts of Shell's extractive industry in Nigeria are massive. A United Nations Environmental Program report on oil spills in Ogoni found that the company operated below international standards, creating massive hydrocarbon pollution, which led to serious groundwater contamination. Oil pollution had specifically affected the vegetation, wetlands, and fish stock over an extensive area. The extent of the pollution made revegetation extremely difficult. The report concludes that the restoration of the area is possible, but will take up to three decades (UNEP, n.d. b, pp. 12–14).

Although the practice has been outlawed in Nigeria since 1984, mass levels of flaring[11] continue to occur along the Niger Delta (Macdonald, 2009). This releases toxic chemicals that have local impacts and exacerbate climate change. Nigeria's greenhouse-gas emissions are higher than all other sources in sub-Saharan Africa

---

[11] Flaring is the use of flare-stacks to burn off natural gas associated with extraction and refining of oil and gas.

combined (Roderick, 2005). The effects of oil activities are detrimental to all aspects of life within Nigeria and the rest of the world.

The experiences of Sudan and Nigeria with foreign corporations drilling for and exporting oil are similar to what sometimes occurs in other extractive industries. Here, corporations employ national military forces or mercenaries from a private military firm to protect lands that are crucial for mining operations, a practice that has been called "militarized commerce" (Kyriakakis, 2007; see also Forcese, 2001). However, there have not been any reports of such practices by foreign agribusiness operations. There have been situations where the governmental forces shot and killed demonstrators against land acquisition, and displacement of local residents to make room for farming for investors (Reuters, 2013a). Reports of displacements rarely mention the number of relocated people, leaving the impression[12] that their numbers are not large. The Gulf states have plenty of complex domestic and/or regional issues (e.g. treatment of foreign labor in the Gulf, status of women in Saudi Arabia, Sunni–Shia tensions) and are likely to steer clear of adding to their complex set of challenges. They conduct their affairs as most capitalists would, guided by their national interests. While the Gulf states are not necessarily compassionate capitalists, they certainly do not want to be viewed as exploitative neocolonialists. This delicate balancing act is work in progress.

Most of the land transactions to date were concluded in opaque circumstances in countries that have a high score on the Corruption Perception Index, yet a few were signed in developed countries where governments are more transparent, accountable to the people, and corruption is generally lower (Transparency International, 2013). Examples of the latter include countries like Brazil, Canada, Poland, and Turkey. The reality is that local impacts of land deals vary greatly depending on the business and political environment in the host country, and on the integrity and business model of investing agencies.

Some aspects of land transactions are controversial, and how these farms will function in times of political or economic turmoil has yet to be tested. Foreign agricultural investors who seek opportunities in developing countries often accept significant risks, which vary from the geopolitical and social to the climatic. When a host government is corrupt, disliked, or distrusted because it lacks popular legitimacy, land transactions may be cancelled by the next government. Transactions that are perceived as being shady and lack popular support may experience local resistance, which could range from peaceful protests to acts of sabotage of farming operations, the most vulnerable of which is likely to be their transportation and irrigation infrastructure. In an apparent acknowledgment of this, the Pakistani

---

[12] If the number of displaced people was in the "thousands," the assumption is that media outlets cannot overlook such a major news story like this.

government's land offer to Gulf investors came with its willingness to "hire a security force of 100,000 to protect the assets" (Economist, 2009). When food prices spike again, would the host countries' security forces use harsh tactics to quell popular efforts to block food shipments out of the country? If they did, would the investing country or agency tolerate such a public-relations debacle?

Risks to agricultural foreign investors could be from the country of origin or the target one. The following focus on Saudi investments in Ethiopia helps illustrate the fragility of the investment-trade environment and the challenges that face Gulf states in operationalizing policy directives with respect to farming abroad.

Ethiopia was one of the first countries that Saudi Arabia targeted for land acquisitions. The number of Saudi agricultural investors in Ethiopia is more than 400, and their activities are focused mostly on the production of wheat, rice, and barely. The quality and extent of the infrastructure varies greatly from one part of the country to another. In certain less-developed areas, some Saudi investors have erected bridges where rivers are numerous, and built roads. Saudi investments have created jobs for Ethiopians where the number of employees in small farms exceeds 1,500, and have trained many on using agricultural equipment and other modern farming methods (Alasmary, 2013). One of the most prominent investors is billionaire Sheikh Mohammed Hussein Ali Al Amoudi, an Ethiopian-born Saudi citizen whose father is Yemeni and whose mother is Ethiopian. He used one of his firms, Saudi Star Agricultural Development, to lease about 10,000 hectares in the northeast of Ethiopia, and another 290,000 hectares of farmland in western Ethiopia were in the process of being leased in 2012. In 2011, he was believed to be the largest absentee landowner of Ethiopian farmland. Saudi Star Agricultural Development now owns 10,000 hectares in Gambella Regional State, plans to add 500,000 hectares, and is already exporting rice from it (Horne, 2011).

Gulf investments like those by Al Amoudi have faced local resistance. For example, employees of Al Amoudi's firm Saudi Star were ambushed in 2012, and as a result, five of them were killed. Although Al Amoudi was lured to Ethiopia, and is considered by some as being too big to fail, the fast changing events may force the company to readjust the focus of its investments. Since 2007, Jenaan, an Abu Dhabi investment firm, has acquired some 67,200 hectares of arable land in Egypt. The company faced labor strikes, diesel shortages for the agricultural machinery, and was faced with $43 a ton in export tax. Jenaan had intended to grow fodder to feed livestock in the UAE, but its losses and new tariffs forced it to change its plans; it now grows wheat for consumption within Egypt (Reuters, 2013a).

Saudi Arabia's Agricultural Development Fund, responsible for facilitating investments abroad, has come under criticism from Saudi agricultural investors in Ethiopia for its rigidity and lack of cooperation. For example, it demands that investors applying for a loan must provide assurances of political stability in the target country, a condition that investors find almost impossible to meet. Investors also complain that, five years after King Abdullah's initiative was announced, there is still no procedure in place to facilitate the exports of produce from farms abroad to Saudi Arabia (Alasmary, 2013). They are concerned about the "regularization" of the status of Ethiopian workers in Saudi Arabia that led to the eviction of tens of thousands in 2013, and how that might affect the public's acceptance of Saudi investments in their country.

The governments of the GCC, save for Qatar, had very warm relations with the regime of Hosni Mubarak, and Egypt was a favored destination for their tourists. This made Egypt attractive to Gulf corporations that had ties with the government, a partner in many of their investments. These companies were courted by the Mubarak regime and invested heavily in the country, especially in real estate. Shortly after the regime was overthrown by millions of street protesters, the new authorities in Cairo started civil and criminal investigations of land transactions of many Gulf investors.

Prince Alwaleed bin Talal, a Saudi businessman and chairman of the Kingdom Holding Company, has invested more than $2 billion in Egypt, much of it in lands that he had purchased during the reign of Mubarak. In 1998, a Kingdom Holding unit called Kingdom Agricultural Development Company (Kadco) purchased for $127 million some 42,470 hectares (420 million square meters) of land in the South Valley Development Project (SVDP) in Toshka to develop vast agricultural activities. The SVDP is commonly referred to as the Toshka project. The mammoth $2 billion project was to create approximately 2.8 million jobs, develop industries and tourist attractions like safari and agrotourism, and new towns and villages where up to six million people would live. The government cost of the all the development projects would reach $90 billion (Warner, 2013). Water supply for the new project was to come from efficiency increases, (brackish) groundwater, and from the Aswan High Dam. When the government of Gamal Abdel Nasser was building the Aswan High Dam, it incorporated a 14-km Toshka Overflow Canal that was designed to protect the integrity of the dam and prevent flooding by channeling excess water from Lake Nasser's western shore to the Toshka Depression. The SVDP was an extension of this preliminary work.

This mega project was an attempt by Mubarak to secure a lasting legacy on par with former Egyptian leaders such as Muhammad Ali and Gamal Abdel Nasser. Given its location deep in the Sahara desert close to the border with Sudan, far away from population centers and public services, and given that the Nile waters are barely sufficient to meet the needs of the people in the basin, Tony Allan of the University of London's School of Oriental and African Studies called the Toshka plan "preposterous" and a "national fantasy" (Gladman, 1997). Mubarak's government was unable to entice many Egyptians to relocate there, nor to secure World Bank funding, so it once again turned to the Gulf states and got them involved in the project. Sheikh Zayed bin Sultan El Nahayan, the late president of the UAE,[13] contributed $100 million for a 72-kilometer-long canal which was named after him.

In 2011, a new transitional government accused the Mubarak regime of "crony capitalism" and jailed many of its ousted leaders. It then froze the assets of the former agriculture minister, Youssef Wali. Shortly after the success of the revolution, the new public prosecutor stated that investigations had revealed that the land transaction undertaken by Kadco had "provisions that violated the law and gave the company unjustified benefits" (Hope, 2011). The company had been illegitimately exempted from fees and taxes for 20 years, and the size of the acquired land was twice the lawful maximum for a government property to be sold. Kingdom Holding reported that it had not forfeited its ownership of the Toshka land nor was the land seized. The land was acquired in accordance with all legal requirements.

Ultimately, the company which had initially considered international arbitration against the Egyptian authorities reached an initial agreement that involved returning 30,000 out of 41,000 acres of agricultural lands (Saigol, 2011), without Kingdom Holding acknowledging any wrongdoing. Kadco's decision to opt for a low-profile, out-of-court resolution keeps the door open for future investments, does not damage investor confidence in the country and company, and saves face for a globally-prominent Arab investment firm. Of all the numerous high-profile assets that Kingdom Holdings owns, the Toshka lands were the most politically sensitive and controversial, which does not bode well for its global image and reputation.

As a result of their frustrations in Africa, some Arab land investors have liquidated their holdings in Ethiopia, and many others are in the process of doing the same (Alasmary, 2013). Furthermore, it was recently reported that corporations based in the Gulf states are rebalancing their investment portfolios in agriculture by focusing on established agro-producers and farmland acquisitions, this time in developed countries (Reuters, 2013a). In 2013, Al Dahra, an agricultural company that is based in the UAE, bought eight agricultural firms for $400 million in Serbia (Reuters, 2013a). People in developed countries are believed to be more receptive to foreigners buying some of their farmland or investing in their agricultural businesses. While the cost of doing business in

---

[13] Despite the fact that Toshka is now judged as a "failed agriculture megaproject", companies like Al Dahra and Jenaan from the UAE are investing in Egypt's southern desert and plan to grow and sell to the Egyptian government several hundred thousand tonnes of wheat. The "low yields, poor soil quality and uncertain water supplies make such a venture seem reckless". Its political value, however, outweigh its economic risks. The rulers of the UAE emerged as staunch supporters of Abdel Fattah al-Sisi, Egypt's President, who promised his people to reclaim desert lands and create jobs (Reuters, 2014c).

developed countries is usually higher than in developing countries, there is no indication that the oil-rich Gulf states were necessarily after "bargains." Also, agricultural investments in developed countries carry low political or social risks, and the business environment is usually predictable.

Foreign governments acquiring extensive land holdings in riparian states where international water rights have not been settled may create tensions between riparians, ones that could escalate to conflicts. One of the first people to link these transactions with water is Peter Brabeck-Letmathe, the chairman of Nestlé, who said the land purchases in poorer countries were about water, a resource that is offered for token sums or for nothing, and described them as "the great water grab" (Economist, 2009). Water is typically implicitly bundled with farmland transactions where it may be in the form of precipitation, soil moisture, ground, or surface water. In some cases, the acquired land falls along a river bank (Pearce, 2012); the choice of the location is not accidental. Here, water is a paramount criterion for investors, and it needs to be either on or in close proximity to the target land. In many cases, investors would need to develop this land by, for example, making certain areas flat, removing large boulders, and setting up an irrigation system where water may have to be piped in. Access to water needs to be factored into the fees or price paid for the land; if it is not, then the water-grabbing label would apply. This, however, assumes that our globalized, interconnected world "is flat" where the locals have more or less equal knowledge and business skills as the international investors; a very ambitious assumption.

Farming abroad for countries that are deficient of renewable water and arable land is a form of water demand management. Demand management is a term used to describe the efficient use of available resources, thereby reducing the need for additional supplies. The Gulf states have been gradually outsourcing parts of their farming sector. Their initial move focuses on limiting areas sown with thirsty, yet low-value crops like wheat, barley, and fodder. This major water-saving policy slows water depletion yet it does not rein in the exorbitant water consumption by residents. The Gulf states need to consider all their water-conservation options in order to achieve sustainable management of their limited natural resources, and they need to respond to the growing criticism of their foreign land acquisition.

In order for the Gulf states to win the hearts and minds of locals in host countries, they should consider a dual-track approach to the management of foreign land holdings where one track would cover the management of large-scale operations that are capital-intensive and mechanized, making for efficient production. Another approach could be based on small-scale tracts where Gulf investors would use labor-intensive technologies that would employ many local workers, and produce perishable products like fruits and vegetables, most of which would be sold locally at competitive prices. Gulf investors should also work with host governments to build the physical infrastructure such as roads and ports, and to bolster human capital that specializes in domestic and international law. These would enhance the trading environment, help enforce contracts, and uphold property rights in the host country. National governments should provide services that lubricate trading such as market data (e.g. trends in crop prices, weather forecasting, etc.), financial credit, communication systems, and facilities for storing harvests so they don't spoil before reaching the market. Finally, they should develop grades and standards for produce, and connect peasant farmers with the markets at home, the wider region, or globally. In doing this, the investors would be involved in grassroots developments that work with the local government to empower the farmer by providing the necessary conditions, as well as the institutional and physical infrastructures that will enable their continued development (loosely based on Christy *et al.*, 2009).

3.6 Conclusions

Since the 1980s, the Gulf states have experienced substantial increases in incomes per capita and quality of life. These have affected dietary habits, which now include many more water-intensive proteins, and have necessitated a large foreign workforce. As a result, the modern-day Gulf states cannot come close to feeding the people from the land. The growing global demand for productive farmland has led economically advantaged states (China, South Korea, Saudi Arabia, and many others) and institutions to acquire farmlands and invest in agribusinesses mostly in developing countries, and especially in Africa. Most of the Gulf states, led by Saudi Arabia and the UAE, have unleashed ambitious programs to boost food production on farmlands located in foreign countries yet controlled by national corporations or by arms of the national governments. Ethiopia and Sudan have been the prime investment targets. They, like other African countries that have attracted foreign agricultural investments, tend to have authoritarian governments, as well as high levels of corruption and poverty. Land transactions in this setting have invited charges of "land grabbing" to which the Gulf states appeared oblivious. To alleviate concerns over unwelcome consequences of this asymmetric relationship between patron and client, the Gulf states should proactively consider all the likely social, environmental, and reputational impacts of land acquisitions.

Just like the notion of "land grabbing" has been the focus of intense debate, so has the question of whether water scarcity will induce cooperation or trigger "water war" between national or sub-national actors (Swain, 2001; Gleick *et al.*, 1994; Amery, 2002). An online news magazine asked well-known water specialists for their understanding and take on this question. Zeitoun (2009) rebuffs the idea of water war stating that "the absence of war does not mean the absence of (water) conflict", which causes much "suffering throughout the world". On the other hand, Gleick (2009) argues that the "water war" question is irrelevant because it depends on how you define "war," and recasts the issue to conclude that there is a definite "connection between freshwater and conflict, including violent conflict". Similarly, Pearce (2009) asserts that water does "contribute(s) to the reason people sometimes go to war". Sandra Postel (2009) cuts through the debate by rightly saying that the phrase "water wars" "has come to encompass a spectrum of conflicts over water. ...the preoccupation with whether an outright war (i.e., military conflict) between nations will erupt over water has overshadowed the larger threats to social stability and human well-being posed by mounting water stress worldwide". To describe agricultural investments as "land grabbing" is counterproductive because it sensationalizes a serious issue that touches the lives of so many people, and polarizes the debate into a "with us" or "against us" dichotomy. Injustices need to be heard and addressed, and host-country locals should have their own voice in the debate.

Large tracts of acquired farmlands in Africa and Asia remained idle years after having been acquired by Gulf states. It appears that these foreign agricultural investors approached the issue of food production through the narrow lens of a commercial transaction ("buy the land"), but did not sufficiently consider the human, ecological, logistical, and political (national and community scales) parameters that they needed to align for operations to start: the land needs to be prepared, and the human and physical infrastructure needs to be in place; qualified people need to sow seeds in suitable ecosystems, and crops need to be harvested, processed, and sold abroad or in the host country. This is an elaborate process that requires intimate knowledge of the host socioeconomic and ecological environment; hence, imported farmers are not likely to succeed because agricultural lands, like all places, are "directly experienced phenomena of the lived-world"; they are neither abstracts, nor concepts (Relph, 1976, p. 141).

Unlike other Gulf countries, Qatar has the most experience operating farms and agribusinesses in developed countries such as Australia. It, for example, learned the value of hiring lobbyists who would occasionally explain to politicians and to the general public the objectives of their farming operations and how they are mutually beneficial. Preparing society to accept your investment is an important part of doing business abroad. Very recently (2013), the Gulf states have embarked upon a qualitatively different investment approach: to acquire farmland or agribusinesses in developed countries. This approach increases business costs but decreases investors' exposure to social or political risks.

The large-scale outsourcing of agricultural activities may shake feelings of national identity and pride. The people of the Gulf have become very urbanized in recent decades, hence they have been removed from traditional farming activities such as pearl diving and land-based agricultural activities. People who farm the land develop an attachment to it, which explains the expression "rural nationalism." The Gulf Arabs, with the possible exception of Omanis, are moving from rural nationalism into what may be called "urban internationalism." The Gulf governments should expand the range of activities that develop and enhance feelings of rootedness, sense of place, pride in the homeland, cultural and other activities, and policies that would help in national identity formation. They should also devise and implement policies that are more embracing of expatriates, and set citizenship criteria that make cultural and demographic sense. Rapid urbanization and the demographic imbalance, where a massive number of foreigners toil the farms and occupy most jobs, all have social, political, and security implications that deserve greater consideration by researchers. In January 2013, the government of the UAE adopted a new law that introduces mandatory military service for men and is voluntary for women, a law that is understood as deepening loyalty and solidarity of the local population (Bayoumy, 2014).

4

Responses to water insecurity

4.1 Introduction

The current global water crisis is largely a result of "profound failures in water governance, i.e., the ways in which individuals and societies have assigned value to, made decisions about, and managed the water resources available to them" (UNDP 2004, p. 2). Some 80 percent of "the world's population is exposed to high levels of threat to water security", a challenge that concerns arid as well as non-arid countries. The same study argues that "massive investment in water technology enables rich nations to offset high stressor levels without remedying their underlying causes, whereas less wealthy nations remain vulnerable" (Vörösmarty *et al.*, 2010, p. 555). Margaret Catley-Carlson of the World Economic Forum stated that new, innovative approaches to water management need to be developed in the next two decades otherwise, "large parts of the world will face a serious and structural threat to economic growth, human well-being, and national security. Some will feel the heat sooner than others" (WEF, 2011, p. xxii). Technological innovations such as desalination are inevitable for those countries and regions that are arid and wealthy enough to afford them. In addition to economic tools, technological and softer, non-technical measures such as education need to be deployed in sync to help develop cultural norms that value freshwater resources and encourage residents to act on and pay for their protection, provision, and sustainable management.

Freshwater availability in the Gulf states is an order of magnitude lower than demand has been in recent decades. It takes one liter of freshwater to produce one calorie of food, hence each person needs an average of 2,800 liters of water a day in order to live a healthy and hygienic life.[1] Given the huge volumes of water that are needed to produce food, and the fact that renewable water in the Gulf states is well below the water poverty index of 500 m^3 per person per year, this chapter

---

[1] The Food and Agriculture Organization of the United Nations uses 2,800 kilocalories (kcal). However, this figure varies depending on whether a person is on protein-rich or vegetarian diet (WWAP, 2009).

106

considers both water and food, and their nexus. In Saudi Arabia, for example, "ground-water reserves have been severely depleted," which adversely affects the agricultural sector and many cities, such as Jeddah (Lidstone, 2010).

The natural resources that the Gulf states are rich in but underutilize are sunshine, desert land, and saltwater; they hold much promise for a sustainable future. While the Middle East and North Africa region has many different sources of renewable energy, solar irradiance is significantly more abundant than others (World Bank, 2012). A focus on solar energy could ensure a continued supply of power to desalination plants, a reliable supply of water to residents, and would enhance environmental sustainability.

Water scarcity occurs when demand is greater than physical availability, and when the infrastructure's capacity to access, control, store, and distribute the resource is curtailed by economic or political considerations. It is also affected by the quality of institutions responsible for providing water services. Water stress, especially if it is severe or protracted, often yields conservation policies that are commonly based on technological, managerial, and cultural considerations, such as the following:

- Installing separate meters for landscaping makes it possible to identify what percentage of water is consumed for food and general hygiene as opposed to water used for esthetic purposes. Usage data would allow managers to decide if they need to make adjustments to water supply or to tariffs charged.
- Maintaining the water infrastructure. Water audits for customers should include leak detection and repairs.
- Conducting technical fixes, such as retrofitting faucets, showerheads, toilets, dishwashers, wash machines, and irrigation systems with low-flow or ultra-low flow replacements.
- Extending wastewater connections to all households in the country.
- Developing and nurturing a culture of conservation through a public information campaign that targets the general public through advertisements, educational curriculum, and through promotion by national leaders, be they cultural, political, religious, or economic icons.
- Developing institutional capacities that reflect environmental realities.
- Enhancing and expanding specialized human capital, and ensuring consistency between policies that emanate from different government agencies yet impacting water resources in one way or another.
- Utilizing conservation pricing in different forms, such as charging consumers the cost of sustainably providing water. It could also include charging a small, or token tariff for a minimal amount of water that the average household uses for their daily activities; above that amount, significantly higher tariffs would kick in (loosely based on Shetty, n.d., and Vickers, 2001).

4.2 Governance

Geoclimatic challenges, rapid demographic growth, high economic growth rates, and the politicization of water provision converge to produce societies that are dependent on desalination and have one of the highest per capita water consumption rates in the world. To meet the astronomical rise in demand for water and electricity, Saudi Arabia needs to spend SR800 billion ($213.3 billion) between 2014 and 2024, where the private sector's share of the investment is expected to be 30 percent in either water or power. This in a country where the demand for water and power is rising at an annual rate of 8 percent (Reuters, 2014b). The lack of surface water supplies has forced people to focus on sub-surface supplies and desalination. Jury and Vaux (2005, pp. 15718–15719) argue that "The era in which most growing demands for water could be addressed by developing water supplies through large-scale infrastructure like dams in canals appears to be over or nearly over." The geography of the Gulf states, however, has forced people of that region to choose desalination; this is the inevitable destiny that nature determined for them. Economies of scale and rising human demand require building ever larger plants, and this has a direct relationship to vulnerability.

Historically, water supply has been the dominant management approach where governments have provided communities and industries with the freshwater they need. It is sometimes called the hard-path approach because supply is typically infrastructure-based. In recent decades, questions have been raised about the efficacy of this approach and whether greater benefit can be obtained from existing allocations. Demand management seeks to create market and non-market incentives, mechanisms, and interventions that would reduce water use through greater efficiency (Boberg, 2005). A World Bank (Parker, 2010, p. xii) report argues that "effective demand management is one of several critical challenges worldwide in the face of increasing water scarcity." It also states that "demand for water can be affected by three broad sets of measures: pricing, quotas, and measures to improve water use efficiency." Brooks (2006) defines demand management as technical, economic, administrative, financial, or social methods that are used to achieve one of more of the following goals:

1. Reduce the quantity or quality of water required to accomplish a specific task.
2. Adjust the nature of the task or the way it is undertaken so that it can be accomplished with less water or with lower quality water.
3. Reduce the loss in quantity or quality of water as it flows from source through use to disposal.
4. Shift the timing of use from peak to off-peak periods.
5. Increase the ability of the water system to continue to serve society during times when water is in short supply (Brooks, 2006, p. 524).

In the water-intensive farming sector, demand management would require greater efficiency and better allocation of available water resources where, for example, farmers:

- practice water-conserving tilling;
- improve irrigation scheduling;
- irrigate at times that are critical to a crop's yield;
- maintain their irrigation system; and
- choose non-thirsty crops.

Some Gulf states have embraced elements of demand management where water education and technology retrofits are being implemented at varying paces. Management deficiencies persist in the areas of human-capital empowerment and raising water tariffs to capture a greater share of the actual cost of providing freshwater to the masses. Water professionals in the Gulf states tend to be expatriates; like all other guest workers, they work at the pleasure of the employer and could be deported to their country of origin with little advance warning. Hence, water professionals are typically unempowered, and their qualifications are sometimes deficient or simply different from those of other expatriates who may come from different countries. These factors make most of these professionals feel timid, vulnerable in their positions, and inclined to generally "go with the flow." This suppresses employees' willingness to propose creative solutions to challenges at work, and reduces their commitment to the workplace.

4.3 Principled pragmatism of cost recovery

A detailed assessment of the economic impacts of water use is a prerequisite to sustainable water management. Wasteful use of water due to government subsidies distorts its market price and dilutes the public's appreciation of its value. However, scientists at Carnegie Mellon University found that to make $1 worth of milled and refined sugar cane, it takes 270 gallons (1,022 liters) of water. The same study showed that $1 worth of animal food requires 200 gallons (7,57 liters) to produce, and $1 worth of grain takes 1,400 gallons (5,299 liters). These data include direct and indirect use of water for the purposes of irrigation, but also for processing, packaging, and shipping of food crops to the supermarket (Blackhurst *et al.*, 2010). Such hydro-economic data are particularly valuable in situations where water scarcity is acute. By knowing the market price of each virtual liter of water, decision makers can then create or revise a water policy, one that yields maximum economic benefits from every unit of water used or saved. One of the recommendations of the International Conference on Water and the Environment in Dublin in 1992 is that water should be treated as an "economic good." While a vigorous debate was triggered about this perspective (Savenije, 2002; Briscoe, 2011), it made it easier for some governments to introduce water fees – especially those that do not charge for water or offer it at a heavily subsidized price.

McPhail *et al.* (2012) suggest that economic instruments that could be deployed to reduce water consumption include charging lower tariffs on water-saving technologies and on non-thirsty crops, as well as restructuring pricing policies in the agricultural sector where farmers pay for the volume of water used, not the area of land irrigated. Another important approach is water cost recovery, whereby a service provider would capture sufficient revenue from customers to cover their current and some of their future costs (Rodriguez *et al.*, 2012). These should include the costs of operating and maintaining the service, capital costs, and also the recovery of asset depreciation and savings for possible expansion of the services the water infrastructure provides.

Although desalination technology is centuries old, desalinated water is more expensive than water from natural sources. Hence, it is only an option for wealthier countries. The planning of a desalination project involves a feasibility study and engineering designs that identify technical specifications which are capable of achieving the outcomes, tendering, and construction. The process takes a minimum of three years before water can be produced. Decision makers have to make long-term forecasts that identify the most-suitable desalination technology to adopt.

The capital costs for desalination plants that use the same technology, feed-water type, and are the same size can vary considerably from $72 million to $307 million (Abazza, 2012 quoting Wittholz, 2007). The commercial life expectancy of a desalination plant can reach 40 years, even though investors amortize their costs over 20 to 25 years. Other parameters to consider in the decision-making process are the cost of operations and maintenance, which include the costs of energy and chemicals, as well as the availability of steam and power (Utilities ME, 2012; Voutchkov, 2012). The costs of mitigating environmental impacts are sometimes considered as well. Furthermore, during the lifetime of a plant, residents' perception of environmental concerns change and costs of assets vary, sometimes widely.

A World Bank report argues that solving problems associated with water scarcity are "considerably more complex and nuanced" than a generalized pricing formula such as getting customers to pay the full cost of water. It introduces the concept of "principled pragmatism" where users are made to "take financial and resource costs into account when using water". The important contribution here pertains to the "pragmatist" side of this approach where "solutions need to be tailored to specific, widely varying natural, cultural, economic and political circumstances, in which the art of reform is the art of the possible" (World Bank, 2004b, p. 22). This tailored approach is helpful due to the myriad implications for

Table 4.1 *Cost of desalinated water* over time*

1980	$4.50–$1.50
2005	$2.0–$0.50
2012	$1–$0.50

* Per cubic meter.
Source: Abazza (2012).

water pricing. Across the Gulf states, political, social, and economic circumstances vary, sometimes considerably.

In recent decades, the cost of producing desalinated water has been decreasing, and sometimes rapidly (Table 4.1). In the United Arab Emirates (UAE), tariffs on desalinated water were $0.06 per cubic meter in 2009; this constituted 0.3 percent of the actual cost of the produced water, which was $1.75 per cubic meter (excluding capital, and fuel). Because about two-thirds of the water delivered is essentially free, the income gained from conservation "would be negligible" (EAD, 2009). The ruler of Dubai, Sheikh Mohammed bin Rashid Al Maktoum, ordered Dubai Electricity and Water Authority (DEWA) to double its "free water quotas allocated for low-income Emiratis in Dubai from 10,000 gallons/month to 20,000 gallons/month for houses and farms. One and half fils[2] will be charged for each gallon consumed above the set limit" (WAM, 2011).

The Sheikh also gave Emiratis in Dubai up to 6,000 kWh of electricity a month per household for free, and explained his decision as "The UAE leadership" wanting to provide nationals "with a prosperous life and services at every house without having to bear additional burdens." In the same statement, he also said "water and electricity are national assets which have to be protected from wasting and therefore preserved for future generations." Mariam Mohammed Khalfan Al Roumi, Minister of Social Affairs, said that some 7,016 low-income Emiratis would benefit from the measure (WAM, 2011). Charging very little or nothing for a critical and scarce resource like water does not contribute to its sustainable management. The executive secretary of the International Sustainable Energy Organization observed the obvious when he said that water in the UAE is "running out because it is being given for free" (Landais, 2007).

Improving cost recovery could be implemented in phases, which would make it more acceptable to consumers. However, an Organization for Economic Co-operation and Development (OECD) report argues that in regions or countries where water tariffs are extremely low compared with the full cost recovery, "a gradual approach may not be sufficient and more drastic action may be called for" (Parris, 2010, p. 19). It is highly unlikely that the Gulf states will, in the short to

---

[2] One Emirati fils is the same as 0.01 dirham. One dirham equals to 0.27 US cents.

medium term, drastically reduce water subsidies. However, it is possible to phase in a policy of full cost recovery of water with expatriate workers. Were this to happen, it would significantly reduce overall consumption, especially in countries where the vast majority of the population are expatriates, such as in Qatar, the UAE, and Kuwait. Some countries have taken drastic measures regarding select subsidies in the agricultural sector, such as banning certain thirsty crops or, as in the case in Saudi Arabia, lifting the subsidies off wheat – a crop with low economic value (see Chapter 3 for details on this issue). The Gulf states' farming abroad, and their growing reliance on virtual water, is both a pragmatic acknowledgment of the hydro-geographic reality and a partial reformation of their agricultural policies. They also need conservation-inducing policies, such as higher water tariffs in the domestic, industrial, and farming sectors.

The emirate of Abu Dhabi initiated a scheme to treat and reuse all waste-water across its jurisdiction (WAM, 2013a), and started inspections of farms under its domain. During these inspections, it discovered cases of transfer and selling of groundwater by some farm owners and tank drivers. This is in violation of Law No. 6, which regulates groundwater wells and associated efforts in the emirate (WAM, 2013b). Illegal pumping of groundwater in Abu Dhabi has become a real problem. The sale of groundwater accelerates the exhaustion of this resource, deteriorates its quality, and boosts salt accumulation in the soil (WAM, 2013b).

In recent years, Oman "curtailed the drilling of wells and legislated against expansion of irrigated land" (Lidstone, 2010). Additional measures that ought to be considered by the Gulf states, include phasing out flood irrigation. Jackson *et al.* (2001) argued that balancing current and future demands for freshwater requires:

(a) sufficient and timely monitoring of water quality, as well as prevention of pollution flows to freshwater sources that support ecological functions and services;
(b) protection of aquifer recharge zones and of watersheds;
(c) laws that treat surface and subsurface waters as an integrated whole;
(d) authorities to have water supply and demand plans well into the future; and
(e) the application of economic tools to induce water conservation and protection of freshwater sources.

While perennial surface-water flows are non-existent in the Arabian Peninsula, groundwater has been polluted by activities such as the application of agricultural chemicals, whereby untreated wastewater or industrial pollutants are released, which leach into and contaminate the mostly non-renewable groundwater supplies. Furthermore, the Gulf states initiated long-term strategic plan in the 1990s, an approach and practice that took off a decade or so later. Countries and major cities started drafting documents to capture their general vision for the future. For example, the government of Qatar published its national development strategy for 2011–2016, which contained a chapter on "sustaining the environment for future generations" (QGSDP, 2011). An even longer-term strategic document was issued by the Government of Abu Dhabi (2008), *The Abu Dhabi Economic Vision 2030*, and one by the Government of Bahrain (2008), *From Regional Pioneer to Global Contender: The Economic Vision 2030 for Bahrain*. The latter states that the three guiding principles for this kingdom's long-term economic vision are sustainability, competitiveness, and fairness.

4.4 Subsidies or sustainability?

Water governance is about "the processes and institutions by which governments, civil society, and the private sector make decisions about how best to use, develop and manage water resources" (UNDP 2004), and how best to deliver water resources to customers. In the Gulf states, civil societies are weak at best, and often non-existent. The political culture is one that does not allow much space for public input nor participation in the decision-making process. Therefore, crafting and implementing water policies tends to follow a top-down approach where questioning the political authorities is uncommon, frowned upon, and can sometimes bring retribution on the "troublemaker."

The total annual pretax energy subsidies in Middle East and North American (MENA) countries was $237 billion in 2011, which is "equivalent to 48 percent of world subsidies, 8.6 percent of regional GDP, or 22 percent of government revenue." The vast majority ($204 billion) of these subsidies were in oil-exporting countries (Sdralevich *et al.*, 2014, p. ix). Subsidies transcend the region where, for example, the European Union and the United States heavily subsidize the farming sector in their respective countries. Many utilities in developing and developed countries, including the United States, are barely able to cover the basic operation and maintenance costs of the water services that they provide. Implementing full cost recovery of operations and maintenance is hardest to capture and implement in irrigation sectors in developing countries. Subsidies, however, are hard to reverse once they are in place. For example, the total value of all exemptions and subsidies offered by the United States "tax code is around $1.3 billion, an amount that could be significantly trimmed without damaging the economy" (Micklethwait and Wooldridge, 2014, p. 132).

Underpricing fuel makes it cheaper to pump groundwater, in turn lowering the cost of irrigation and discouraging conservation and cultivation of value-added crops. This explains why groundwater extraction exceeds annual recharge, causing "groundwater depletion that represents an estimated wealth loss of as much as 1–2 percent of GDP for some countries" (Sdralevich *et al.*, 2014, pp. 25–26). Wasteful water consumption which results from low tariffs can "cause the depletion of water resources" and raise the future costs of production, "as water can only be found by drilling at greater depths or conveying from longer distances" (Rodriguez *et al.*, 2012).

The low water and energy tariffs that are pervasive in the Gulf states make it commercially unattractive for homes and corporations to install sophisticated, water-efficient technologies. Therefore, per capita water consumption in the Gulf states, as well as the energy and carbon footprints of produced water, are higher than in Western countries. Given that oil and gas are exhaustible natural resources, the Gulf states' reliance on such rent will eventually taper down making heavy water subsidies unsustainable. Countries of the OECD adopted a policy of "sustainable cost recovery" (Parris, 2010, p. 19). It suggests moving the water sector onto a financially sustainable ground where a country finds "the right mix between the ultimate revenues for the water sector . . . tariffs, taxes and transfers." These "3Ts" need to be carefully considered. Each country must find "its own balance" among these key sources of finance (Parris, 2010, p. 19). Policies that are inspired by this experience could help put the Gulf states on a sustainable water-security path.

Kuwait, like other Gulf states, has a comprehensive welfare system where citizens and expatriates benefit from generous state subsidies on power and other utilities. In 1966, tariffs that were levied on electricity bottomed out at 2 Kuwaiti fils[3] per kilowatt-hour for regular consumers, and 1 fils per kilowatt-hour for industries. These were the prices that the government charged customers, rates that were "very low even by regional standards". Almost five decades later, the same 1966 rates were still in effect in 2014, even though electricity tariffs for luxurious homes were raised to 10 Emirati fils (Fattouh and Mahadeva, 2014, p. 9). In addition to the super-low tariffs, the central government routinely absorbs them. In 2005, each Kuwaiti household was relieved from paying up to 2,000 dinars ($6,895) of their power bill, and the government rarely penalized those who didn't pay. Kuwaitis (1.1 million) and expatriates (2.35 million) pay 2 fils a kilowatt-hour (MacDonald, 2010). As for the water sector, while the government has a pricing schedule, "water bills are not collected" (Milutinovic, 2006). The daily, per capita water consumption is 500 liters in Kuwait and Qatar (Fattouh and Mahadeva, 2014; Varghese, 2013), 550 liters in the UAE (National, 2014), and 265 liters in Saudi Arabia (Tago, 2014). The consumption of Kuwait and Qatar is more than double that of the world's average. These data are particularly anomalous because the Gulf states have some of world's lowest levels of renewable freshwater resources per capita.

Saudi Arabia is consuming around 2.8 million barrels of oil per day, which is over one-quarter of its total oil production. If consumption is not curbed, the

---

[3] One Kuwait fils is the same as 0.1 dinar. One dinar is equal to 3.39 US cents.

kingdom is projected to become a net oil importer in 2038 (Lahn and Stevens, 2011). Of that consumption, it burns a total of 1.5 million barrels per day to power desalination plants so to produce freshwater. A report by the National Commercial Bank, a regional investment bank, argues that "poor management, population growth and the promotion of energy and water-intensive lifestyles have pushed the Gulf states into a crisis" (Lidstone, 2010).

In the last few years, Gulf leaders have taken note of, and some have become concerned about, their subjects' disproportionate level of consumption of oil in their cars, electrical power in their homes (especially for cooling), and the fact that they are among the highest consumers of water per capita in the world. Decision makers are treating this issue with greater levels of seriousness so much so that boosting water security is identified as a strategic objective, and a senior UAE official said that the importance of water sometimes "supersedes the importance of oil" (Khaleej Times, 2012). KISR, a government-backed research center in Kuwait, found that, as of 2011, the cost of providing freshwater "exceeds 1.2 $ billion annually". If Kuwait continues with its business-as-usual approach, a government study estimates "that by the year 2050, given current consumption patterns, the majority of the country's revenue that is generated by oil will be required to fund the increased production of desalinated water" (Fattouh and Mahadeva, 2014, p. 7). This financially unsustainable reality is getting the attention of national governments in the region. Some officials are suggesting targeted and phased introduction of a tariff structure.

In the fall of 2013, it was reported that the government of Kuwait plans to review subsidies on goods and services. They are bleeding state coffers to the tune of $15.9 billion a year (Reuters 2013b; Utilities ME, 2013a). For the 2014 fiscal year, subsidies are expected to cost 5.1 billion dinars ($18.2 billion) for a population of 3.8 million (Reuters, 2014a). In Bahrain, the governor of its Central Bank, Rasheed al-Maraj, said that the government wants to curb the excessive use of energy by designing a new subsidies policy, one that targets the needy in society (Utilities ME, 2013b). *The Economic Vision 2030* for the kingdom asserts that "subsidies for water, electricity, gasoline and food ... will exclusively target the needy to reduce costs and avoid over-consumption of scarce resources" (Government of Bahrain, 2008, p. 19). A member of Bahrain's parliament wrote that his country's gross domestic product is $30 billion and yet subsidy programs cost the government an "alarming" $3.4 billion in 2013, "which is more than 11 per cent of the country's GDP, something quite out of the ordinary by global standards" (Ali, 2014).

The UAE Minister of Energy said that his country wants to reduce domestic consumption by targeting energy subsidies to locals, hence getting expatriates to pay more. The UAE currently pays about 85 percent of the power and water production costs for local citizens, and 50 percent for foreign workers. The International Monetary Fund (IMF) estimates that subsidies amount to 5.5 percent of GDP in 2012 of the UAE (Said, 2014). At the single emirate level, and with conservation as a goal, the Dubai Electricity and Water Authority (DEWA) reconfigured its price structure to charge higher fees to consumers who exceed a certain minimum level (Landais, 2008). The flat fee was 3 fils per gallon. Under the new structure, each gallon will continue to cost 3 fils but only for the first 6,000 gallons. Beyond that, the cost will be 3.5 fils per gallon, and 4 fils for those who exceed 12,001 gallons. Furthermore, a similar sliding scale structure was introduced for electricity fees. The new price scale, which went into effect in the spring of 2008, impacts 20 per cent of the city's consumers, mostly commercial and some residential ones; native residents are exempted from it (Landais, 2008; Thompson, 2008). However, in nearby Abu Dhabi, the Environment Agency (EAD) there declares that "groundwater is free in Abu Dhabi" (EAD, 2009, p. 52); sub-surface water is primarily used for irrigation in the emirate and the country (EAD, 2009). And, given that the agricultural sector is the largest consumer of water in the country and in other Gulf states, the current policies encourage wasteful uses.

Governmental subsidies of essential goods and services are part of the implicit social contract that Gulf monarchs have made with their citizens. The latter have a sense of "entitlement" to the natural wealth in their country, and have, therefore, grown deeply dependent on subsidies for water, energy, and for other social services. Furthermore, this "sharing" of natural wealth earned ruling families a certain degree of legitimacy and acquiescence from the masses. Consequently, subsidies are politically difficult to change; charging market price for water is more delicate than for energy because it is considered a human right and a basic human need. The problem with subsidies is that they tend to be "regressive" where the wealthiest people benefit most; they have larger houses and bigger cars, and engage in more energy-intensive activities. An IMF study reported that subsidies are an inept method of helping the poor, squeeze government coffers, and distort relative prices in the economy, which spurs overconsumption. Energy subsidies reduce the level of hydrocarbon exports and the revenue associated with them (Sdralevich *et al.*, 2014); a major concern for governments whose economies are anchored on exporting such natural resources.

Sustainable water development is about managing current human uses by enjoying the numerous range of benefits associated with water while also ensuring that future generations will have sufficient amounts that are of adequate quality (Mathews, 2005; Richter, 2010). Therefore, sustainability requires a long-term perspective where water production, use, reuse, and disposal of untreated volumes have low social and environmental impacts, and rest on on sound budgetary practices. The potential threats that wars, oil spills from tankers, and drilling operations pose to water intakes of the large number of desalination plants that dot the western shores of the Persian Gulf make it harder for central governments to pursue environmental sustainability. The realization that water quality in the Persian Gulf is beyond the control of central governments led them to decide jointly to build desalination facilities on the Arabian Sea, from which freshwater would be piped to various Gulf states.

Water service providers need to be financially sustainable if they are to be reliable into the future. This goal is achieved through cost recovery, whereby a service provider is able to meet the costs of operations and maintenance, renewal of current infrastructure, future expansion of existing facilities, and capital costs. Cost recovery would help water-supply firms become more financially sustainable and embrace water conservation. A water industry leader in the Gulf states said that the people of that region can "afford to pay the market rate" for water (EIU, 2010, p. 14). Decision makers in the Gulf have been reluctant to craft such policies where citizens pay market prices for natural resources because they place a high premium on political stability and therefore avoid politically contentious policies that may generate a boomerang effect. Affected stakeholders may leverage their political capital against the state by structuring their protests, as well as by collaborating and sharing information and experiences to achieve their goals (den Hond and de Bakker, 2012; Keck and Sikkink, 1998).

Since the 1990s, the Gulf Cooperation Council (GCC) countries have started taking steps towards sustainable development, the pace of which took time to pick up speed. Qatar's *National Development Strategy 2011–2016* seeks "to improve the management of water resources" to meet the needs of future generations (QGSDP, 2011, p. 220). A larger effort is being undertaken by the Abu Dhabi Urban Planning Council (UPC), which developed the institution of Estidama (the Arabic word for sustainability) for "promoting thoughtful and responsible development . . . while recognizing that the unique cultural, climatic and economic development needs of the region require a more localized definition of sustainability." For this emirate, the goal of Estidama is "to preserve and enrich Abu Dhabi's physical and cultural identity, while creating an always improving quality of life for its residents on four equal pillars of sustainability: environmental, economic, social, and cultural" (Estidama, n.d.).

Achieving a sustainable standard of living would require the people of the Gulf states to reflect on and (re)consider how they design and build, live, farm, and how they source things. In recent years, they have become serious about efficiency and conservation. This gradual change in attitude is being helped by technological improvements. In one decade, the performance of the multi-effect distillation (MED) process moved from producing 7 kg of distillate water per kg of steam to 9–10 kg of freshwater (Utilities ME, 2012).

4.5 Technological interventions

In addition to the economic tools that are discussed in the previous section, infrastructure, science, and technology have an important role to play in enhancing water security, conservation, and overall sustainable management of the resource. Plugging leaky pipes moves these states a step closer towards sustainable water management. Globally, 30 to 35 percent of all freshwater supplies don't reach the customers because of leakages; hence, the GCC states fall within the global average of unaccounted-for water (UFW). The UAE has the most efficient distribution system with 13 percent losses whereas those for Saudi Arabia and Qatar are around 35 percent (Saraf, 2013). Between 1993 and 2001, Bahrain managed to reduce the rate of UFW or leaks from around 32 percent to slightly over 23 percent (GWI, 2003). A decade later, the rate was still at 20–23 percent, even though the government's goal is to reduce it to about 15 percent (Saraf, 2013).

A report by the government of Qatar estimates that the UFW is around double that in Western countries. This and leakage from sewage systems add to the cost of construction, adversely affect marine life, increases salinity in groundwater and reduces agricultural yields (QGSDP, 2011). In addition to these economic costs, desalinated water remains expensive because many of the plants are over ten years old, and rely on technologies that are less efficient than modern ones. A study by the Center for Clean Water and Clean Energy estimates the cost of UFW in Saudi Arabia to be $820,000 annually (quoted by Saraf, 2013; also Al-Zahrani and Baig, 2011; Avancena, 2010; Lidstone, 2010; Oxford Business Group (OBG), 2008), while Qatar's loss is around $300 million per year (QGSDP, 2011). Saudi Arabia loses 1.1 million cubic meters a day of water (Maree, 2008) which is on par with the daily production of the Ras al-Khair desalination plant on the Persian Gulf, the largest in the world. This $7.2 billion plant started operating in the spring of 2014. The city of Riyadh wants to reduce its UFW from 35 per cent to 7 per cent by 2028, which could nudge city leaders to raise "water tariffs sooner than previously expected" (Maree, 2008).

The growing fear of water-supply disruption has spurred Gulf governments to initiate strategic storage for emergencies. When it announced the project, the UAE government bypassed any reference to security concerns, but mentioned that stored water is an insurance against oil spills or plant breakdowns (Reuters, 2010). In Kuwait, a technical breakdown led to the shutdown of three state-owned oil refineries when maintenance work on a transformer caused a power blackout in the whole country. This failure cut off power to four of the five huge power-generation and water-desalination plants in Kuwait (Cassidy, 2004). While this breakdown lasted for only a few hours, it illustrates the nexus between water and energy, the linkage between their infrastructures, and the fact that water and energy security are intertwined.

The water-related physical infrastructure is receiving greater levels of protection against physical and cyber attacks. Desalination plants are fortified by high fences, sophisticated electronic security monitoring, and armed forces. To visit a plant in any Gulf country, one is required to go through elaborate security background checks. Prince Turki Al-Faisal (2011) stated that Saudi Arabia is hardening its oil-producing infrastructure by investing billions of dollars on security and surveillance systems, and by creating (in 2005) a specialized 35,000-strong force to protect all energy installations against internal and external threats. Members of this force enlist in extensive technical training programs that benefit from the expertise of American advisors.

Corporations that are based in the Gulf and beyond have been trying to convince Gulf leaders to adopt ultra-modern forms of farming. Agricel, a venture capital firm, has soilless farming technology which claims to deliver water savings of up to 90 percent, and increase food yields by up to 50 percent. The company's chair is Omar Ghobash, the UAE's ambassador to Russia (Kerr, 2012). Agricel aims to commercialize the research of Professor Yuichi Mori of Waseda University in Japan. Qatar seeks to achieve significant levels of food self-sufficiency by using state-of-the-art technologies, and by relying on solar energy to desalinate water for irrigation. Once this goal is met, farmers will be asked to stop using aquifer water altogether. Qatar's aquifers are severely depleted, and the country is starting to view them as a strategic reserve (Fuchs, 2012). In 2008, Saudi Arabia restructured its water management and launched an efficiency initiative which lowered consumption by as much as 30 percent (Thompson, 2008). This was also the year the kingdom decided to halt its subsidy for wheat production and to abandon the decades-old policy of wheat self-sufficiency.

The UAE's Minister of Environment and Water, Rashid Ahmed bin Fahd, developed a federal law requiring rationing of water and energy consumption (Khaleej Times, 2012). One of the objectives of this ministry's strategic plan is to reduce water consumption per capita from 360 to 200 liters per day, a rate that is closer to international averages (Khaleej Times, 2012). The country is aggressively attacking high water consumption in Abu Dhabi. It planned to retrofit 55,000 homes and 5,000 public buildings with a device that can reduce the flow of water from 12 to 3 liters a minute. Also, some 300 mosques have already been outfitted with the device (AlRumaithi, 2010). In collaboration with DEWA and Sesam Business Consultants, Grohe initiated a Green Mosque initiative in Dubai in 2009. It installed new self-closing taps and low-flushing toilets in the Abu-Hamed Ghazali Mosque. To optimize efficiency, the new technology accounted for the average time, water pressure, and amount needed for the pre-prayer ablution. DEWA compared water consumption after the mixers were installed with the same period 12 months earlier and found that average monthly consumption dropped by around 30 percent. Grohe then installed its water-saving fittings in Khalifa Al Tajer Mosque in the UAE. The initiative is embraced by many mosques in Saudi Arabia, Bahrain, and other countries in the region (Nair, 2013).

4.6 Limits of technology

Complex technologies like those found in desalination plants fail. Natural hazards and political instability can disrupt the production and supply of freshwater, and natural conditions vary immensely from year to year, potentially leaving communities and countries in a state of severe drought. Such issues have forced the people of the Middle East, and especially those in the Arabian Peninsula, to build massive structures to harvest what they can of the meager water flows that are available to them. They built the Great Dam of Ma'rib, which used to block water flow in the valley of Dhana in southern Yemen. Ruins of this ancient dam date back to the eighth century BCE. As pressure mounted on Kuwait's groundwater resources in the early twentieth century, the government established the "Water Company." Created in 1939, the Water Company used a fleet of 40 small ships to import freshwater from Basra to Kuwait city, and also built several reservoirs scattered around the capital to store freshwater. In the current era, surface storage in the Gulf region is sometimes described as "impracticable or impossible" (EAD, 2009, p. 186). This, however, pertains to supplies in times of emergencies for the region's huge population (compared to the middle of the last century). The available strategic reserves in the region are sufficient for "mere days" (EAD, 2009, p. 186), not weeks or months.

Irrespective of their civilizations or levels of economic development, people living in arid areas have historically devised methods that are appropriate to their circumstances in order to manage their meager water resources. This has varied from saving grain from one year to the next, to building terraces that increase soil moisture, and erecting reservoirs and large dams; all these techniques made communities more resilient and better prepared to tolerate unpredictable environmental adversities (WWAP, 2009). For example, irrigation's contribution to the production of reliable harvests and food security is growing. Around 2005, Saudi Arabia, like other Gulf governments, started investing heavily in building strategic water reservoirs, especially for densely populated urban centers. Their stated objectives are to (Al-Hilali, 2014):

- reduce pressure on the water-supply system;
- provide water for emergencies; and
- ensure the people's water security.

Saudi Arabia is planning to create different strategic reserves to meet the needs of every city for seven to ten days. In 2013, it allocated 51 billion riyals (or $13.6 billion) to create strategic water reservoirs across the country (Asshaykhi, 2013). The province of Jeddah invested 9 billion riyals ($2.4 billion) in constructing reservoirs with the capacity of one or two million cubic meters each for a cumulative capacity of ten million cubic meters. They are to be completed by 2015 (Al-Hilali, 2014). The city of Jeddah's share would be six million cubic meters of freshwater. In the UAE, had desalination plants ceased operations in 2010, the country would have had four days of freshwater. In the fall of that same year, it started building a massive underground reservoir to be filled with 26 million cubic meters of desalinated water. This is billed to be the world's largest underground reservoir, which will store some 90 days of rationed water for the country's population (Reuters, 2010).

According to a recent report (Utilities ME, 2014a), Qatar's General Electricity and Water Corporation (Kahramaa) wants to provide residents with seven days of potable water that will be stored in five huge new reservoirs based at Umm Birka, Umm Slal, Al Thumama, Rawdhat Rashid, and Abu Nakhla. When completed in 2017, the strategically important reservoirs will be among the biggest in the world in their category. Existing and planned reservoirs will have the capacity to store 17 million cubic meters of potable water. They will have associated pumping stations, and a network of large-diameter ring pipelines that will link individual sites together. In addition to these, existing and planned secondary reservoirs will help Qatar preserve its water security. A Qatari official said that the project is hugely significant for his country and is "sure to have an impact around the world where other countries are seeking to ensure strategic water supplies for their populations" (Utilities ME, 2014a). In 2011, the government of Qatar decided to allocate $2.75 billion to build an above-ground reservoir that can store 1.9 billion gallons (7.2 billion liters) of water. Above-ground structures are expensive and less-efficient when compared to subterranean locations. Underground storage reservoirs require minimal maintenance, are cheaper to store water in, difficult to pollute, are not prone to evaporation problems, and are not susceptible to breakage or cracks like above-ground structures (Chatila, 2011).

Finally, the GCC countries agreed to homogenize criteria for water from desalination and for water transport. In addition to this linked water infrastructure, they also agreed to build a giant desalination plant on the Omani coast of the Arabian Sea, and to invest $7–10.5 billion to build a common water-supply network. They also want to build huge desalination plants in Oman that will be linked through pipelines that move water to member states, and to construct facilities for strategic storage (Reuters, 2013a; Sophia, 2014). The advantages of the Arabian Sea are that it is far less polluted than the

Persian Gulf, and is less susceptible to terrorist threats and to major spills, be they oil or radioactive material.

4.7 Desalination using alternative energy

Saudi Arabia's National Science Agency is pursuing solar-based desalination plants, with goals of reducing the cost of providing energy and freshwater by 40 percent and of lowering the environmental impacts of that process. The first solar-powered desalination plant with a capacity of 30,000 cubic meters per day was established in 2012 in Al-Khafji on the east coast. It serves 100,000 people and relies on new nanotechnology that was developed by the King Abdulaziz City of Science and Technology in association with IBM (Ghafour, 2010).

Saudi Arabia receives an average of nine hours of sunlight per day, which translates into more than 3,000 hours per year (Zeigler, 2013; Chaoul, 2013). In comparison, Germany relies significantly on solar energy[4] despite receiving 500 hours of sunlight per year (Chaoul, 2013). Saudi Arabia has ambitious plans for the future of its solar energy sector; its finance minister, Ibrahim Al-Assaf, and its Petroleum Minister, Ali Al-Naimi, have both said that the kingdom is working to become a net exporter of renewable energy, especially of solar power (Amiri, 2010; Lahn and Stevens, 2011). In addition to investments in solar energy, a former diplomat and Saudi Royal family insider, Prince Turki Al-Faisal (2011) eastimates that by around the year 2020, nuclear power will play a major role in the kingdom's energy mix. The government is investing more than $100 billion to build at least 16 nuclear power plants. Both sources will help meet the rising domestic energy needs without significantly reducing its oil exporting capability (Friedman, 2011). This optimistic prediction is not shared by independent experts. For example, Krane (2012) quotes Saudi and British studies which show that the kingdom will consume its total production capacity of 12.5 million barrels of oil a day by 2043 or sooner turning the kingdom into a net oil importer.[5]

The UAE is building four nuclear energy plants in the Barakah region, the first of which will start operations in 2017. When all plants become commercially operational in 2020, they are expected to provide "up to a quarter of the nation's electricity needs. Nuclear energy will also save the UAE up to 12 million tons in greenhouse gas emissions every year" (Utilities ME, 2014c). The national nuclear program has allowed the UAE to consider using this source of energy for

---

[4] Solar and wind made up some 17 percent of energy generated in Germany, and solar energy generation alone grew by 28 percent in the first half of 2014 compared to the same period a year earlier (Renewables International, 2014).

[5] These forecasts assume that Saudi Arabia would not be able to boost its production above a maximum of 13 million barrels a day, and would continue its very generous energy subsidies, and hence wasteful consumption.

desalination (Dziuba, 2011). Emirates Nuclear Energy Corporation (ENEC) announced that around 1000 UAE companies have contracts totaling $1.7 billion for which they will provide various services and products to support the development of the country's nuclear energy plants, and of a "local nuclear energy industry supply chain" (Utilities ME, 2014b). The management of water and energy sources needs to be inclusive of diverse stakeholders.

The growing number of Arabs calling for a voice in their governance has a direct bearing on the management of their natural resources. Orr *et al.* (2009) argue that better water management has to happen at the "local scale if benefits are to be felt across different sectors." Furthermore, a United Nations assessment of world water issues made it clear that the global water crisis "is one of water governance, essentially caused by the ways in which we mismanage water" (WWAP, 2003, p. 4). This highlights the roles that education, public awareness, and good governance play in enhancing water security. The EAD in Abu Dhabi distributes an easy-to-read manual to farmers that explains and illustrates the various techniques that can be used to conserve water. The manual is published in the official national language, and also in other south Asian languages, as well as English (EAD, n.d.). This amounts to an official nod from the UAE to the new reality of farming in the Gulf states as a whole: farming has become the domain of foreign workers, and the farm now functions as a destination for a weekend retreat for the family of local citizens. This is strongly evident in Kuwait, Qatar, UAE, and Bahrain and to a lesser degree in Oman and Saudi Arabia.

The drying effects of climate change, government subsidies of freshwater supply, and the rapid increase in water use per capita are triggering revisions in water policy. Some Gulf states have made radical decisions with respect to their agricultural sectors by deciding that their agriculture is too small and too economically and socially feeble to save. A report by the OECD argues that in "extreme cases", drought-prone areas should "consider abandoning agriculture completely" (Parris, 2010, p. 23). Kuwait would be a good candidate for this; it is a country that virtually ran out of groundwater in the 1970s. The UAE and Bahrain would make for candidates that ought to reflect on the efficacy of making the desert bloom. It is worth noting that fishing and other sea-based economic activities are far more central to Gulf culture and tradition than irrigated land-based farming.

4.8 Conclusions

The Booz & Company consulting firm studied water issues in the Gulf and wrote that "for years, inefficiencies in the GCC water sector have been overlooked." It argued that if the Gulf governments promoted water conservation through tariff reform, education, use of water-efficient fixtures, they could "reduce domestic use from 250 litres per person daily to as little as 190" (Fayad *et al.*, 2011). The UAE's Federal Electricity and Water Authority (FEWA) Strategic Plan for 2014–16 aims to reduce consumption of water and electricity by:

- establishing a culture of conservation in society with special attention paid to students and mothers;
- reaching agreements with government agencies to implement best practices; and
- reconsidering the tariff for electricity and water service (Emirates247, 2013b).

Even with public awareness campaigns, a culture of water conservation is not likely to take root and have a real impact on society unless it is "accompanied by changes to regulations or pricing" (EIU, 2010, p. 14). This explains why despite the various efforts of the Gulf states, "public attitudes towards energy and water conservation – including curtailment of subsidies – remain resistant to change" (EIU, 2010, p. 2). Technological fixes, which make a real difference in saving water, are likely to remain the most prominent approach to conservation because they sidestep the politically delicate issue of getting residents to bear a much larger proportion of the financial burden of water and energy.

When scientists and decision makers work in tandem, they can make farming choices and draft policies that are dynamic and sensitive to local conditions. For example, flowers and fodders such as the African daisy and beach evening primrose can survive in very saline conditions. This is important because "more than 25 percent of irrigated land globally is affected by salinity and an additional 15 percent is waterlogged" (Drummond, 2011). In the Gulf states, (saline) *sabkhas*[6] are common, and groundwater is often brackish.

Water security is enhanced by the sustainable management of the Gulf states' resources, and by an interdisciplinary approach where social, technical, and economic tools are deployed together in a coordinated fashion. This is a particularly hard path to take in the Gulf states. Two, potentially three, generations of nationals have been raised with extensive social welfare systems, energy that is heavily subsidized, and water resources that are practically free of charge. Unlike technological changes, cultural and attitudinal changes evolve slowly, and take years if not decades to develop.

---

[6] Sabkha is a salt flat that forms along arid coastlines like the UAE, Kuwait, and others, which border the Persian Gulf.

5

The future of water and food security

5.1 Introduction

Water scarcity may have adverse effects that include undoing social peace, further disturbing and degrading "natural" systems by encouraging the use of low-quality water for irrigation, and may result in greater aquifer contamination. Suppressed crop yields can result from such practices, and farmers may react by removing marginal lands from cultivation. Disturbing and degrading natural systems also increases the likelihood of irreparable damage to aquifers and soils over time, and the possibility of triggering water conflicts (Al-Jamal and Schiffler, 2009, p. 481) (Figure 5.1). The confluence of environmental, geopolitical, economic, and demographic factors made desalination technology the destiny for the people of the Gulf states. Hydrocarbon wealth facilitated this hydrological independence, but it comes with its own set of risks. This volume argues that the Gulf states face many threats to their water and food security, which require a comprehensive approach, one that embodies technological, political, as well as socioeconomic modifications to current policies that influence foreign workers and water management. It is the socioeconomic reforms that are likely to be the most transformative, and hence controversial.

Despite the popular image of the Arabian Peninsula as a desolate landscape of endless sand dunes, the region once contained a large number of oases, as well as isolated and sometimes contiguous areas with fertile soils. In Oman, most people are not nomads but "live in towns and villages along the wadi-beds", and agriculture has the potential to prosper "as irrigation is nearly everywhere possible" (Zwemer, 1907, p. 601). Writing in 1907, Zwemer (1907, p. 603) observes that eastern Arabia up to Qatif, near Dammam, "could be a veritable paradise under a stable government. It is a land of streams and fountains, subterranean but inexhaustible; and even now, with primitive schemes of irrigation, has wide fields of rice and wheat and extensive date-orchards." In 1951, Douglas D. Crary, an assistant professor of geography at the University of Michigan, Ann Arbor,

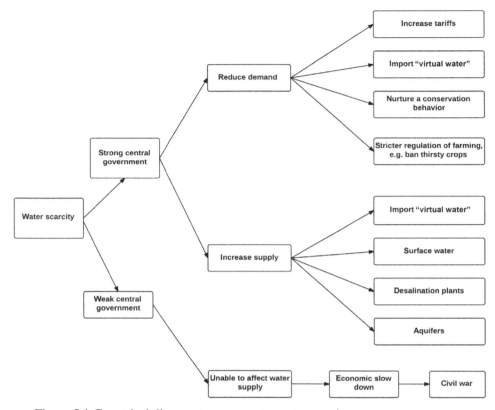

Figure 5.1 Countries' divergent responses to water scarcity

lamented the popular media's misrepresentation of the Arab Gulf. Crary rejected the media portrayal of the Gulf as a place where a "little American know-how" and a "handful of Texas farmers" who, quoting an American official, taught "the Arabs how to farm" have produced "a land of plenty" that made even the "the sands of Arabia . . . bloom again". The author then states, "practically everything Americans know about the basic principles of irrigated agriculture was known to the Arabs two thousand years ago." Since the first millennium BCE, Arabs and other natives living in the Fertile Crescent and western Arabian Peninsula had developed very complex methods for capturing, channeling, and using runoff (Sherratta, 1980). Crary (1951, pp. 366–367) credits Americans with the ability to "apply mechanized power, the use of fertilizers, seed selection, and other new knowledge to an age-old agricultural system and thus point the way to increased yields and somewhat greater self-sufficiency. This, in essence, is the objective of the current Saudi Arabian agricultural program."

An agricultural development project planned "along Western lines" was started in 1937 at Al Kharj, around 100 kilometers southeast of Riyadh. For this program, the Saudi Arabian government imported pumps and mechanics from Egypt and Iraq, a farmer from Palestine, and purchased six tractors. The project took place under the supervision of farmers from Qasim. "It is an interesting fact that this first use of power-driven farm machinery in Saudi Arabia took place before Americans had anything to do with the agricultural program" (Crary, 1951, p. 367). It is also interesting that the Saudi government at that time relied on the science, engineering, and farming skills of fellow Arabs, something that has become much less common in recent decades.

Crary (1951, p. 383) astutely observes "the critical and most frightening aspect of the expansion of agriculture in Saudi Arabia is the relationship between mechanical power and the potential water supply." He adds that the development of large-scale agricultural farms increases the pressure on water supplies and observes that this phenomenon is so common that, in some arable areas, an "enormous amount of power has recently been applied to obtaining water from the ground, and already ground-water levels have sunk." Saudi Arabia then started to have power-operated drilling rigs and diesel-driven pumps "delivering a thousand gallons of water a minute" while donkey wells could only deliver two or three gallons in the same amount of time (Crary 1951, p. 383). Technological diffusion brought the mechanical pump and modern drilling techniques to the Saudis in the 1950s and helped them, especially the farmers, enjoy better lives and more steady incomes through secure yields from irrigated farming. However, this very technology had an almost immediate adverse effect on the water table. Crary (1951, p. 383) concludes his study by warning, "should the water supply fail, Saudi Arabia would return to the desolation whence it came." This 60-year-old observation, while made before the oil boom had transformed the Gulf state, holds true today. At that time, the water supply could have failed if groundwater was exhausted beyond the reach of residents or if the water pump was to break down and could not be repaired or replaced. Today, the continued supply of freshwater from mega desalination plants is central to peace and stability in the Gulf states.

The Gulf states were incentivized to revive farming abroad as a result of regional instability, the politicization of food and trade (e.g. sanctions), and food export restrictions that were imposed by a few countries during the 2008 global food crisis. In the mid to late 1970s, the Arab world was pinning its food security hopes on the massive area of arable land of their fellow Arab country of Sudan. Of this area, only some 10 percent was under cultivation. The United Nations named Sudan (along with Canada and Australia) as being able to help "counteract the world food shortage". Arab countries, especially Saudi Arabia and Kuwait, invested large sums of money to develop Sudan's farmland. The government of

Sudan agreed to undertake 100 projects that would, in the span of ten years, supply 40 percent of the Middle East's food imports at a cost that exceeded $6 billion (Kaikati, 1980, p.122). Subsequently, coups in Sudan and regional political developments undermined that agreement and the food plan was abandoned. Since 2008, the Gulf states' food security efforts included agricultural investments in Sudan, but went far beyond to include many other countries in Africa and on other continents. Qatar chose the slightly different path of investing in some agricultural corporations in developed countries, and more so in farming-related scientific and technological innovations in an attempt to have some degree of food self-sufficiency. The Gulf states most recent water–food configuration is similar to the behavior of autarkic states that strategically prioritize their efforts at establishing greater levels of self-sufficiency through a certain degree of vertical integration (i.e. acquisition of farmland abroad), and national water-supply systems (i.e. desalination). This effort of partial withdrawal or relative disengagement from the international trade system is intended to offer the Gulf states greater assurances that their food and water-supply venues are more secure.

5.2 Science, engineering, and national security

In 1957, B. K. Blount, British chemist and government administrator of scientific and engineering research, argued that the "decisive factor in the politics of our age is science and its practical applications in engineering projects of all kinds" (Sprout, 1963, p. 205). A country's "systematic application of scientific knowledge and engineering techniques" will help it build its political power and influence. This is dependent on strategic variables, the first being the number of highest-caliber scientists relative to other nations; the creative and bold thinkers who would propose daring projects, and have the capacity to convince politicians and policy makers of their ideas (Sprout, 1963, p. 205). The second variable is the presence of a much larger number of scientists and technologists of ordinary caliber. Finally, the degree of awareness among decision makers of the key role that scientists and new scientific knowledge could play in enhancing the future of the country and their readiness to seek the advice of scientists is also crucial (Sprout, 1963, p. 206). The last point is particularly challenging because if decision makers are not open to research and development, they tend to "underestimate the rate of increase of scientific knowledge and its engineering application" (Sprout, 1963, p. 193).

As noted elsewhere in this book, the Gulf states' ambitious development projects required them to import guest workers, including professionals, to run many of the specialized infrastructures and to oversee complex engineering projects. However, if a terrorist group in the Gulf states were to behave like the so-called Islamic State in Syria and Iraq in 2014, by beheading foreign workers, host countries would likely experience a significant exodus of expatriates. This could jeopardize the business continuity in vital sectors like oil production and refining, operation of desalination plants, wastewater treatment plants, and the like. Such issues have been on the minds of decision makers who have been sending a growing number of locals to study abroad and specialize in fields related to the continued functioning of their countries' critical infrastructure. This is a realistic option for Saudi Arabia but not for Qatar; it would be a very ambitious goal for the United Arab Emirates (UAE) and other Gulf states to work towards.

In 2011, King Saud University in Riyadh, Saudi Arabia, initiated the region's first graduate-level degree program in desalination (Alawsat, 2011). Oman hosts the Middle East Desalination Research Center (www.medrc.org), a think tank with a regional scope. Politics, however, has gotten in its way. Given that its creation is related to the yet-unsettled Arab–Israeli conflict, most Gulf states are unenthusiastic towards this internationally funded center. On the other hand, King Saud University actively works with international institutions such as the World Bank, and with various arms of the United Nations such as Food and Agriculture Organization (FAO) and the United Nations Development Program (UNDP) on desalination and related environmental issues. For instance, in 2012 the United Nations Educational, Scientific, and Cultural Organization (UNESCO) funded a Water Desalination Research Chair position at Saudi Arabia's King Saud University. This position will focus on planning for the kingdom's water desalination needs in the medium and long terms, and on recommending the most suitable locations and technologies for expanding existing facilities or establishing new ones. The problem is, however, that a very high percentage of nationals leave school at a young age, and that humanities, including theology, continue to attract a large number of majors. Despite its small population size (one million locals), the Dubai-based Knowledge and Human Development Authority reports that some 25 percent of Emirati boys fail to complete high school (Hamdan, 2013).

Rentier states create national and regional programs that distribute revenues to citizens from natural resources, an act that blurs the distinction between public and the private spheres. This distributive model is evident in the extensive social welfare programs that the Gulf states have for their nationals. In exchange for access to these services, governments expect political acquiescence. Such institutional catering creates a "rentier mentality" (Beblawi, 1987; Moore and Salloukh, 2007). Because of this, Niblock (2007, pp. 1–2) argues that the government of Saudi Arabia uses its natural resource wealth to coddle the local population who are "living in a cocoon created by apparently unearned income, divorced from the problems facing other peoples", creating a mindset which "sets a population apart from the global community – creating attitudes and mentalities out of touch with international realities." The country, he concludes, needs to develop a non-hydro-carbon-based economy.

This rentier mentality is strongest among wealthier countries like Qatar and the UAE, and is relatively weak among poorer ones like Bahrain and Oman. While it is entrenched in Saudi Arabia, it is slowly changing. The existing extensive government programs include subsidies for utilities and water for irrigation, a practice that tends to be quite expensive. Between 1980 and 2005, the Saudi government spent about $85 billion on subsidies for wheat farmers (Jones, 2010). Change in Saudi outlook, as mentioned in earlier chapters, can be gleaned by the government's decision to essentially end the wheat-growing program by removing irrigation subsidy for the crop. The new fiscal realities that emerged after oil prices started their downward trend in the fall of 2014 will likely force governments to reduce certain subsidies, especially those that impact guest workers or industries (and largely spare locals).

In 2006, and in an effort to conserve groundwater resources, the government of the UAE phased out its support for the cultivation of the water-thirsty Rhodes grass, the most favored animal fodder in the country, and decided to import different types of hay, which it made available to farmers at subsidized prices (MAF and ICBA, 2012). Gulf News (2012) reported that Rhodes grass consumed more than 59 per cent of the irrigation water used by the emirate of Abu Dhabi each year. And, of the 16,000 farms in this emirate, over 10,000 have already stopped cultivating the grass. Also, an increasing number of farms are installing modern, more efficient irrigation technologies. Buffel grass is replacing some of the Rhodes grass, which means that fodder farmers will use up to 80% less water, depending on the efficiency of irrigation technology that is being used. Buffel grass species, which tolerates poor soil, is indigenous to the Arabian Peninsula and is used as fodder in various countries around the world (Detrie, 2011). Also, the UAE is exploring the use of *Distichlis spicata*, also known as seashore saltgrass, and *Sporobolus virginicus*, plants that can be used as animal forage or in golf courses, and can thrive on moderately salty water (Drummond, 2011; MAF and ICBA, 2012). Oman is also taking steps to conserve water, especially in the agricultural sector (MAF and ICBA, 2012). The policy changes that Saudi Arabia, the UAE, and Oman have put in place are indicators of the fact that decision makers in the Gulf states have developed a more ecologically-friendly attitude towards water resources management, one that also serves their strategic and security needs. It is important to note that these measures don't require sacrifices of citizens as the vast majority[1] of them are not directly affected.

---

[1] For example, very few Gulf citizens actually live or work on farms.

The cost of desalinating a unit of water has dropped steadily over the years. Between the 1960s and the 2000s, the cost of multi-stage flash (MSF) distillation desalination technology has decreased by an average of 44 percent every ten years (Alkaraghouli *et al.*, 2009). In the future, this impressive price trend is not likely to continue – at least not at the same pace. Global economic downturns and growth in alternative energy sources may put downward pressure on oil prices, impacting the Gulf states' ability to maintain their high and growing reliance on energy-intensive desalination. This, as well as the finite nature of hydrocarbon resources and Iran's controversial nuclear program, is leading some Gulf states to initiate nuclear programs for purposes that include energy generation and desalination.

Historically, MSF desalination plants were dominant in the Middle East, and the majority of those plants remain operational today. Ten years ago, a World Bank Report (2004a, p. 58) found that the MSF "process is well understood, reliable and has served the Gulf states well. It has given the Gulf states security of supply." The contemporary security challenge, however, is that many MSF plants have "increasingly taken advantage of economies of scale" (Alkaraghouli *et al.*, 2009). This could pose a security threat because if a mega desalination plant were to fail or break down for an extended period of time, it would create significant social dislocation and political instability.

One of the real threats to water security in the region is a radioactive spill into the waters of the Persian Gulf. This may occur as a result of a natural disaster – for example, the Bushehr nuclear plan is in an active seismic zone – or due to possible military confrontations with Iran regarding its nuclear program. The Carnegie Endowment and the Federation of American Scientists, two American think tanks, described the Bushehr reactor's location at the intersection of three tectonic plates as "ominous". They also reported that "Iran is the only nuclear state that is not a signatory to the Convention on Nuclear Safety, and its nuclear materials and stockpiles are some of the least secure in the world" (Vaez and Sadjadpour, 2013). In the spring of 2013, a 6.3-magnitude earthquake struck some 90 km southeast of the port of Bushehr, where it flattened small villages, killed 37 people, and injured hundreds more. The head of the Islamic state's Atomic Energy Organization said that the quake had "no impact" on the Bushehr nuclear facility, and that it has been "designed to withstand earthquakes of more than 8.0 on the Richter scale" (Torbati, 2013). This was barely reassuring to the Gulf states whose relationship with Iran is one of mutual distrust, verging on animosity.

The security dilemma for the Gulf states is that many of them have taken concrete steps towards the development and use of nuclear energy. About four decades ago, Abdul Fattah (1978, p. 183) argued "slow and careful introduction of nuclear energy in Saudi Arabia is practical" because it would provide for economic diversification in the country. He added, however, that the research agenda should include an investigation of the feasibility of cleaner, alternative sources, like solar and geothermal, to power desalination plants. Subsequently, Kutbi and Al Suliman (1994) found that nuclear desalination is appropriate for Saudi Arabia because it is cheaper and cleaner than fossil fuel. Currently, a new report by the World Nuclear Association (2014) posits "small and medium-sized nuclear reactors are suitable for desalination, often with cogeneration of electricity using low-pressure steam from the turbine and hot seawater feed from the final cooling system. The main opportunities for nuclear plants have been identified as the 80–100,000 m³/day and 200–500,000 m³/day ranges."

The eventual depletion of hydrocarbon resources and the current extremely high per capita consumption of oil make it easy to understand why the Gulf states are pursuing the nuclear option. In the very least, it can be viewed as an energy insurance policy for future generations. However, these states have pursued the path of least resistance by simply "pouring money" on the problem of wasteful consumption. Like water, mandated energy conservation is a politically delicate policy approach because locals have come to view themselves as part-owners of their countries' energy resources. For leaders, resource subsidies are financial "carrots" that help keep the population pliant and governable. These sociopolitical factors have painted the Gulf states into a corner, leading them to turn to the nuclear option as a reasonable path forward. However, given the instability in the region, transnational terror networks, and the social embers in some Gulf states, nuclear desalination technology exposes these states to unprecedented risks that will likely create new and significant political fissures in a fragile social setting.

Gulf leaders are not oblivious to the above-mentioned risks. Their governments are taking some major and serious steps towards the development and use of renewable energy sources. Bassi *et al.* (2010, p.736) argue that because of the nexus between water and energy, greater reliance on renewable energy sources such as wind and solar will mitigate greenhouse-gas emissions and water pollution from power-generation plants, and "can lead to potential reduction in consumptive water use." They add that the development of renewable technologies "in the energy sector may also contribute to the sustainable management of water resources." For example, alternative and renewable energy sources require significantly less water to produce per kilowatt hour than nuclear or hydrocarbon sources (Table 5.1). The Gulf states need to confront their dynamic water problems, and integrated water resources management needs to include the following perspectives (after the WEF, 2009):

1. Science and engineering offer insights and skills that deal with the balancing of water demand and supply, reliability and delivery, and conservation.
2. Economics provides measures that help in managing the economic and financial resources that are needed to implement the projects prescribed by water scientists and engineers.

Table 5.1 *Liters of water used per kWh of electricity generation from different energy sources in the United States*

Energy source	Liter of water/kWh
Nuclear	2.3
Coal	1.9
Oil	1.6
Combined cycle	0.95
Photovoltaic panels	0.11
Wind	0.004

From by Bassi *et al.* (2010, p. 736).

3. Political economy speaks to the incentives, institutions, and water governance issues that need to be considered, and the economy-wide trade-offs that may be needed to achieve different policy objectives.
4. Institutions that are well-conceived and structured are needed to enable the delivery of solutions developed by technical and economic perspectives.

Furthermore, effective and sustainable management of water, especially in arid and hyper-arid countries, is an important component of the national security calculus. Water security should be perpetually informed by carefully targeted research into risks and uncertainties, and a country's management approach should be regularly recalibrated to benefit from increased understanding of socioeconomic, political, and environmental conditions at home, in the region, and abroad. Sustainable water management infrastructure needs to be flexible to respond to new information gained from experiences at the community, national, regional, and international levels (Figure 5.2). This clearly requires an open political system that welcomes input from all stakeholders – a political climate that is largely lacking in Gulf states, except for Kuwait (Freedom House, 2014). Management of scarce water resources is "seen as a cornerstone of national security" and the water infrastructure is "regarded as a public good and receives financing from the national budget" (Jagannathan, 2009, pp. 40–41).

Water management planning has many potential benefits such as:

- more reliable water supply, even during droughts;
- better use of existing water supplies;
- higher crop yields;
- reduced operating costs;
- water-conservation pricing;
- higher revenues for water providers;
- farmers switching to crops that are less thirsty and have a higher market value;

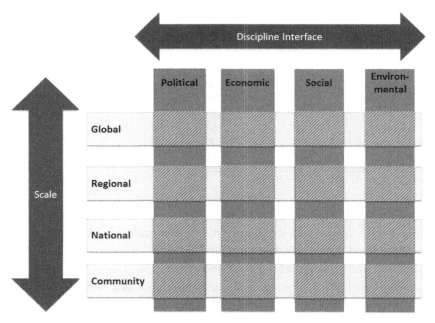

Figure 5.2 Interdisciplinary perspective on water and food security at varying geographic scales

- better management of groundwater's quality and abstraction levels;
- diminished and delayed need for developing new water sources; and
- sustained campaigns to raise the awareness of the public.

Investments in conservation are cheaper and more worthwhile than investments in yet another desalination plant. Conservation, especially when it includes an educational component that is culturally sensitive, will gradually change attitudes and social norms towards how and when water is used. Abu Dhabi has been soft-peddling "estidama" (sustainability) principles by encouraging new privately or publicly owned buildings, be they residential or commercial, to install water-saving technologies. They, for example, install "estidama-compliant" faucets that have a flow rate of 6 liters per minute, which is impressive when compared with conventional faucets whose flow rate is 20 liters per minute (Abdulkader, 2011). What is significant here is the emirate's effort to give sustainability a local flavor that resonates with local culture. Estidama comes from the root words *dama* and *dawm* whose meaning is associated with water. One of the meanings of the former is "continuous rain." The word *deema* is derived from dama which means gentle, continuous rainfall that is free of lightning or thunder (Baheth.info). That is to say, the chosen Arabic word for sustainability, estidama, should resonate with Arabic speakers, especially with those in the Gulf states, because of water challenges that they face. Moreover, they are embracing the idea of sustainability at many different levels, the benefits of which will be long-lasting – affecting successive generations. The discrepancy is that the Gulf states encourage, but don't legislate, water conservation. They take politically safe measures such as mandating water-saving measures in public mosques and government buildings, and lifting subsidies off very few water-intensive, low-value crops, but avoid banning flood irrigation and similar wasteful practices.

The Arab world has a growing number of environmental organizations that have sprouted over the last few decades. For example, the Arab Forum for Environment and Development (afedonline.org) is a multi-national non-governmental organization that is active in conducting studies in the region, raising people's awareness on natural environmental issues, and in starting new initiatives. Cairo is home to an extensive Arabic language sustainability magazine (estidamh.com) that covers environmental news and events from across the Arab world and was founded by the Arab Union for Sustainable Development and Environment (AUSDE). In addition to the magazine, it has a sustainability radio network. There are also environmental organizations throughout the Gulf states such as the Saudi Environmental Organization, which is based in King Abdul Aziz University and works closely with the central government in Riyadh.

In addition to desalination, the Gulf states are pursuing a multitude of other technological approaches to help manage their limited water resources. Water-saving technologies for irrigation or domestic use are gradually being embraced. Soil moisture sensors measure the amount of moisture in the soil and tailor irrigation schedules based on plants' needs. Shutoff devices which turn off irrigation systems when they sense rainfall are in use in some areas. Farms are beginning to integrate sprinkler heads that are more efficient in delivering water and thereby can reduce water loss due to evaporation or wind. Drip or micro–irrigation systems that "deliver low volumes of water directly to plants' roots, minimizing losses to wind, runoff, evaporation, or overspray" are also in use (EPA, 2014). Amongst these technologies, the last is the most efficient yet it is limited by the cost of the infrastructure, and of applying it on uneven or steep terrain.

Arab countries initiated cloud seeding in order to increase precipitation in the 1950s. They resorted to science and technology to modify the weather so as to enhance precipitation. They mostly used aircraft, and in some cases helicopters and ground generators, to inject seeding agents such as silver iodide (AgI) in the winter season, from October to April (Al-Fenadi, 2007). According to Al-Fenadi (2007), the earliest cloud-seeding efforts were in collaboration with Western countries. The French colonial government conducted Algeria's first cloud seeding in 1952, and Libya's was in collaboration with the California-based Sierra Research Corporation. It started in 1971, shortly after Muammar Qaddafi came to power in Libya. The Gulf states were late (Table 5.2) to embrace this technology, with the first

Table 5.2 *The first year of cloud seeding in Arab countries*

Country	Starting year
Algeria	1952
Libya	1971
Morocco	1983
Jordan	1986
Iraq	1989
Saudi Arabia	1990
Syria	1991
UAE	2000

From Al-Fenadi (2007).

cloud-seeding efforts being in Saudi Arabia, followed by the United Arab Emirates (UAE) (Al-Fenadi, 2007). A report by the United Nations Environmental Program (UNEP, n.d. a) states: "The estimated additional rainfall attributable to cloud seeding ranged between 7–16 percent of the natural annual rainfall" and "the use of this technology is in an experimental stage in the other Arab countries, but generally is exhibiting encouraging results."

5.3 People-centered security

In 1904, Leopold S. Amery disagreed with a prominent political geographer, H. J. Mackinder (1904) about his criteria for a globally dominant power. Projecting forward, Amery argued that location and geographical attributes are losing their importance while those who "have the industrial power and the power of invention and of science will be able to defeat all others" (Wilkinson, 1904, p. 441). In the twenty-first century and the information age, the challenge is less one of technology and know-how and more one of social organization and institutional building. Much of the necessary knowledge is readily available, and in other cases can be obtained at a price that the Gulf states can afford.

Some 100 years ago, foreign companies led the exploration and discovery of oil in the Gulf states. These initial steps towards globalization radically transformed every aspect of life in that region. The sleepy communities felt invaded by the cultures and ideas that waves of foreign workers brought with them; locals eventually introduced stringent rules that severely limited their rights and period of employment, even though some found ways to stay for decades. Globalization is the gradual, continuous "expansion of interaction processes, forms of organization, and forms of cooperation" across regions and continents. This "spatial re-organization of production, industry, finance ... causes local decisions to have global repercussions and daily life to be affected by global events" (Cha, 2000, p. 392).

Rosenau (2003) argues that in a globalized world, there is an organic tendency for individuals and societies to integrate across boundaries, and that tendencies towards splintering are steered by local resistance to transboundary activities. In other words, globalization is integrative boundary-broadening while its counterforce, localization, is boundary-heightening, which hinders interactions (Rosenau, 2006). The intense and sometimes novel interactions of globalization processes have generated new challenges that require innovative and unconventional thinking.

In this new security environment "the state remains central, but no longer dominates either as the exclusive referent object or as the principle embodiment of threat" (Buzan, 1997, p. 11). The grievances that guest workers and some Shia natives pose structural challenges to the state system, requiring a break with static policies of the past and the exploration of innovative pathways forward.

Adverse events like natural hazards or acts of terrorism result in corrosive, culturally determined fear that does not lend itself to objective, technical responses. Such responses are likely to separate people and make them more suspicious of one another. People's responses and reactions to adversity and threats are informed by their cultural norms and "behavioral formulae which have come to be part of their everyday behavior and thought" (Giddens, 1991, p. 44). People respond to disasters in a way that corresponds to their own perceptions and priorities, potentially explaining "why they do not do what 'we' want them to do" (Cannon, 2008, p. 356); in other words, "they will behave 'culturally' in ways that do not seem to fit" with forecasted norms of rationality (Cannon and Muller-Mahn, 2010, p. 625).

Durodie (2004, p. 266) argues that "Real resilience requires bringing people together with a sense of common purpose." This is important because in emergencies, the general public are the "real first responders" and as such "it is vital that they be fully integrated into, and engaged by, a set of broader social aims and values." People develop trust and a sense of purpose through "active, political engagement in society." This primarily political, long-term project "cannot be short-circuited by technical means or information campaigns" (Durodie, 2004, p. 266). For instance, technological disasters that were once widely interpreted as Acts of God are now associated with human irresponsibility or malice; the latter are believed to have multiple corrosive effects such as increasing anxiety among the people, as well as aggravating economic problems and social conflicts (Furedi, 2007). This differentiation is especially poignant in the Gulf states where some locals and most unskilled guest workers do not have relationships of trust with the government. They feel a certain degree of cultural subjugation is inflicted upon them by the hegemonic power of the native population. Furthermore, robust confidence in the system is necessary at the commercial level because an organic relationship between national security and business firms would help prevent an extended disruption of certain services. Prevention of a major interruption of water supply, as well as response and recovery efforts, would require operators of desalination plants to have an extraordinary level of coordination with the local authorities and the people themselves.

While financial constraints have been almost non-existent in the Gulf states in recent decades, a time will come when bottom lines become important. The increasing world supply of oil has helped depress prices in the latter half of 2014 to levels that have not been seen in six years. Oil prices are not expected to rebound until at least 2016. Also, in recent decades, Bahrain's reliance on oil revenue has been significantly curtailed as its reserves have become depleted. All hydrocarbon-dependent countries are vulnerable to cyclical commodity prices and to declining reserves. Similarly, given that attacks on the infrastructure are rare, some threats may be difficult to discern and preempt. A study published by the United States' National Academy of Science (Auerswald *et al.*, 2005) argued that "sustaining watchfulness and the ability to deal with low-probability, high-impact events is the single most difficult policy issue facing critical infrastructure providers and homeland security agencies today." It added that protecting critical infrastructure requires the development of "organizational antibodies of reliability that enable society and its constituent parts to be more resilient and robust in the face of new, dynamic, and uncertain threats". This poses a particular challenge for Gulf governments; national security goes far beyond deploying an endoskeleton network of spies among nationals and expatriates, and tougher exoskeletons around critical infrastructures. Security measures that are primarily anchored in technology offer partial, perhaps even temporary, solutions to the national security threats that face the Gulf states; these are the main measures that they have been pursuing. This, however, avoids the more complex, controversial, and time-consuming challenge of recasting the social structures and attitudes of nationals towards "others" in society.

In Israel, it is near-impossible for Palestinian-Israelis to work in "'strategic' state industries such as energy and water" (Economist, 2008a). While Palestinian-Israelis are ostensibly equal before the law, their Christian and Islamic religious beliefs have in effect disqualified them from working in jobs that affect the Jewish state's critical infrastructure. While Israel has qualified Israeli Jews to fill such "strategic" positions, the Gulf states do not have alternative choices. They have long feared that some of their Shia population with ties to Iran may carry out acts of violence and sabotage.

5.4 Conclusions

There is a discrepancy in the Gulf states' attitudes to water and food security. Many of them have taken steps to reduce their dependence on food imports by acquiring farmland abroad, but are not perturbed by dependence on imported desalination technology. As Leopold S. Amery observed in the early 1900s, scientific prowess and inventions endow countries with significant power. The Gulf states have used their strategic resource, oil, to produce water in the homeland. This has enhanced their water and national security, and in turn has helped boost their quality of life and political stability. The technology, however, is mostly imported from Western countries, which could use it as a political leverage to pressure a Gulf government if relations between them were to deteriorate. They need to domesticate their desalination technologies by generating their own scientific and engineering knowledge that has local and regional relevance and application. Gulf societies need to find a suitable formulation that will harness local and other scientific knowledge to advance and entrench sustainable water security in the region. Over five decades ago, it was noted that the geographic distribution of technology and its evolution over time is a "strategic political variable" (Sprout 1963, p. 193).

At the sociopolitical level, the Gulf states do not allow foreign workers to settle permanently in their midst; they want to protect their national identities and cultures. This policy, however, carries elements of risk that the relevant literature on the subject has not considered before. People sometimes act on their grievances, and when they do, they may target a country's (energy) infrastructure. Furthermore, in times of natural or manmade disasters or emergencies, the response to fear is culturally dictated; there are numerous cultural groups in the Gulf states that are deliberately isolated from the local culture, and the majority of the low-income, unskilled workers speak neither Arabic nor English. They may be the first respondents in times of emergencies. The national security of the Gulf states is better served when foreign workers are given political rights and are integrated into the larger culture; this is a vital step towards forging societies that are cohesive and content, with residents who are committed to the welfare of the countries in which they live.

References

Abazza, H. (2012), *Economic Considerations for Supplying Water Through Desalination in South Mediterranean Countries*, Sustainable Water Integrated Management – Support Mechanism (SWIM – SM), The European Union.

Abdul-Fattah, A. F. (1978), Nuclear desalination for Saudi Arabia: an appraisal, *Desalination*, 25, pp.163–185.

Abdulkader, B. (2011), New Abu Dhabi buildings get water saving taps, *Gulf News*, December 21.

Abebe, S. G. (2014), Saudi Arabia's treatment of Ethiopians has been shameful, *The Montreal Gazette*, January 8.

Abed, G. T. (1991), The Palestinians and the Gulf crisis, *Journal of Palestine Studies*, 20, pp. 29–42.

Acton, J.M. and W. Q. Bowen (2010). Civilian nuclear power in the Middle East: the technical requirements, In, Sokolski, H. (ed.), *Nuclear Power's Global Expansion: Weighing its Costs and Risks*. Carlisle, PN, The Strategic Studies Institute Publications Office, United States Army War College, pp. 423–575.

Afrol News (2012), African leaders in Ethiopia land grab, *Afrol News* (Lesotho), January 28.

Agence France-Presse (2008a), Police break up Sharjah riot, *Emirates247*, March 19.

Agence France-Presse (2008b), Gulf states look to harvest food from investment in Asia, *Inquirer.net*, July 20.

Agunias, D. R. (2012), *Regulating Private Recruitment in the Asia–Middle East Labor Migration Corridor*, Bangkok and Washington, DC, International Organization for Migration and Migration Policy Institute.

Akhtar, S. (2011), Food security in the Arab world: price volatility and vulnerabilities and the world bank response, First Arab Development Symposium, Food and Water Security in the Arab World, Kuwait, March 14–15.

Akkad, M. (2013), Qatar's first entertainment complex for "bachelors" to open this fall, *Doha News*, May 26.

Alasmary, S. (2013), Saudis seek to liquidate their agricultural investments in Ethiopia, *Al-Hayat (London)*, December 9, in Arabic.

Alawsat, A. (2011), Water desalination confronts the challenges of human capital deficiencies by launching a master's degree program, UK, In Arabic, http://classic. aawsat.com/details.asp?section=43andissueno=12047andarticle=650920and feature=#. VDm08fldVUU

AlBahli, M. (2013), Reality of Arabic language, *Al-Itihad (UAE)*, March 8, in Arabic.

Al-Faisal, T. (2011), Tour D'Horizon, Royal Elcano Institute, Madrid, Spain, September 26, Saudi-US Relations Information Service (SUSRIS), see http://www.susris.com.

Al-Fenadi, Y. (2007), Cloud seeding experiments in Arab countries: history and results, Ninth WMO Scientific Conference on Weather Modification, Antalya, Turkey, October 22, Geneva, World Meteorological Organization.

Al-Hilali, M. (2014), National water: strategic storage provides 10 million cubic meters extra for Jeddah, *Al Eqtisadiah (Saudi Arabia)*, March 12.

Al-Hajri, K. R. and L. A. Al-Misned (1994). Water resources in the GCC countries: A strategic option. *Renewable Energy*, 5, pp. 524–528.

Ali, J. (2014), The imbalance in Bahrain's generous subsidies, *Gulf News (UAE)*, April 18.

Al Jaberi, S. (2007), *The Implications on the Gulf States of any American Military Operation Against Iran*, Carlisle Barracks, PN, US Army War College.

Al-Jamal, K. and M. Schiffler (2009), Desalination opportunities and challenges in the Middle East and North Africa region, In Jagannathan, N. V., A. S. Mohamed, and A. Kremer (eds.), *Water in the Arab World: Management Perspectives and Innovations, The International Bank of Reconstruction and Development*, Washington, DC, The World Bank.

Alkaraghouli, A., D. Rene, and L. L. Kazmerski, (2009), Solar and wind opportunities for water desalination in the Arab regions, *Renewable and Sustainable Energy Reviews*, 13, pp. 2397–2407.

Alpen Capital (2011), GCC Food Industry, Alpen Capital, June 28, see http://www.alpencapital.com/downloads/GCC_Food_Industry_Report_June_2011.pdf.

Alpen Capital (2013), GCC Food Industry, Alpen Capital, May 1, see http://www.alpencapital.com/downloads/GCC_Food_Industry_Report_May_2013.pdf.

Al Qassemi, S. S. (2013), Give expats an opportunity to earn UAE citizenship, *Gulf News*, September 22.

Al-Rasheed, M. (2011), Sectarianism as counter-revolution: Saudi responses to the Arab Spring, *Studies in Ethnicity and Nationalism*, 11, pp. 513–526.

Al-Rashed, M. F. and M. M. Sherif (2000), Water resources in the GCC countries: an overview, *Water Resources Management*, 14, pp. 59–75.

Al-Rodhan, K. R. (2006), *The Impact of the Abqaiq Attack on Saudi Energy Security*, Washington, DC, Center for Strategic and International Studies.

AlRumaithi, S. (2010), 60,000 Abu Dhabi buildings to get water saving device, *Topnews*, January 14.

Al-Samdan, A. A. (n.d.), *The State of Kuwait's Security Policy – Facing Future Challenges Head-on*, Canadian Forces College, NSSC 6, see http://www.cfc.forces.gc.ca/259/281/276/samden.pdf.

Al-Shaibany, S. (2010), Cyclone Phet damages may cost $780 mln, *Reuters*, June 7.

Al-Shaqsi, S. Z. (2011), *Emergency Management in The Arabian Peninsula: A Case Study from the Sultanate of Oman*. University of Otago, New Zealand: Dunedin School of Medicine.

Al-Zahrani, K. H. and M. B. Baig (2011), Water in the Kingdom of Saudi Arabia: sustainable management options, *The Journal of Animal and Plant Sciences*, 21, pp. 601–604.

Ambah, F. S. (2004), Saudi attacks shake expat workers, *The Christian Science Monitor*, May 4.

Amery, H.A. (2002), Water wars in the Middle East: a looming threat. *The Geographical Journal*, 168, pp. 313–323.

Amiri, A. (2010), Saudi Arabia to build solar energy based desalination plants, *Top News Arab Emirates*, January 27.

Amnesty International (2013), Qatar: end corporate exploitation of migrant construction workers, Amnesty International, November 17.

Amunwa, B. (2011), New research reveals Shell paid militants who destroyed Nigerian towns, Remember Saro-Wiwa, October 23, see http://remembersarowiwa.com/new-research-reveals-shell-paid-militants-who-destroyed-nigerian-towns/.

AQUASTAT (2009), Qatar: Water Report 34, see http://www.fao.org/nr/water/aquastat/countries_regions/qat/index.stm

Arab Water Council (2009), *Arab Countries Regional Report, 5th World Water Forum*, Istanbul, Arab Water Council, February 24.

Asal, V., K. Deloughery, and R. D. King (2013). *Understanding Lone-actor Terrorism: A Comparative Analysis with Violent Hate Crimes and Group-based Terrorism. Final Report to the Resilient Systems Division, Science and Technology Directorate, US Department of Homeland Security*. College Park, MD: START.

Asshaykhi, O. (2013), Al-Muslim: 51 billion riyals for strategic water reserves in the kingdom. *Al Madina (Saudi Arabia)*, December 5.

Associated Press (2004), Oil workers to leave Saudi Arabia, *Washington Post*, May 4.

Associated Press (2012), Israel used "calorie count" to limit Gaza food during blockade, critics claim, *The Guardian*, October 17.

Atiyyah, H. S. (1996), Expatriate acculturation in Arab Gulf countries, *Journal of Management Development*, vol. 15, p.37–47.

Auerswald, P., L. M. Branscomb, T. M. LaPorte, and E. Michel-Kerjan (2005), The challenge of protecting critical infrastructure, *Issues in Science and Technology*, Fall.

Avancena, J. (2010), Millions lost in water leakage per year, says study, *Saudi Gazette*, September 1.

Baheth.info (n.d.), Dictionaries of Arabic language.

Baldwin-Edwards, M. (2011), Labor immigration and labor markets in the GCC countries: national patterns and trends, *Kuwait Program on Development, Governance and Globalization in the Gulf States*, March, no. 15.

Barrett, R. (2011), How a broken social contract sparked Bahrain protests, *The Christian Science Monitor*, February 21.

Barry, C. (2008), Australia's long drought withering wheat, rice supplies, *National Geographic News*, May 29.

Bassi, A. M., Z. Tan, and S. Goss, (2010), An integrated assessment of investments towards global water sustainability, *Water*, 2, pp. 726–741.

Batty, D. (2013), Conditions for Abu Dhabi's migrant workers "shame the west", *The Observer*, December 21.

Bayoumy, Y. (2014), With eye on troubled region, UAE plans military service for men, *Reuters*, January 19.

Bazza, M. (2005), Policies for water management and food security under water-scarcity conditions: the case of GCC countries. 7th Gulf Water Conference, Water Science and Technology Association, Kuwait, November 19–23.

BBC (2012), Analysis: land grab or development opportunity? *BBC News*, February 21.

BBC (2013), Bahrain police attacked in Sitra and Janabiya, *BBC News*, July 7.

Beblawi, H. (1987), The rentier state in the Arab world, *Arab Studies Quarterly*, 9, pp. 383–398.

Benn, A. (2006), US backs Israel on aid for humanitarian groups, not Hamas, *Haaretz (Israel)*, February 16.

Berazneva, J. and Lee, D. R. (2013), Explaining the African food riots of 2007–2008: an empirical analysis, *Food Policy*, 39, April, pp. 28–39.

Berger, E. (2013), Land 'grab' realities, perceptions vary markedly – researcher, *Thomson Reuters Foundation*, July 17.

Berman, S. (2009), What to read on modernization theory, *Foreign Affairs*, March 12.

Birnbaum, M. (2011), Saudi Arabia calm on planned "Day of Rage," but protests spark violence elsewhere, *Washington Post*, March 11.

Bitar, Z. (2009), Emiratisation: working partners. *Gulf News (UAE)*, January 29.

Blackhurst, M., C. Hendrickson, and J. S. I. Vidal (2010), Direct and indirect water withdrawals for US industrial sectors, *Environmental Science and Technology*, 44, pp. 2126–2130.

BMZ (2010), *The Water Security Nexus: Challenges and Opportunities for Development Cooperation. Study commissioned by the German Federal Ministry for Economic Cooperation and Development (BMZ).* Berlin, International Water Policy and Infrastructure Programme.

Boberg, J. (2005), *Liquid Assets: How Demographic Changes and Water Management Policies Affect Freshwater Resources.* Santa Monica, CA, RAND.

Boele, R., H. Fabig, and D. Wheeler (2001), Shell, Nigeria and the Ogoni. A case study in unsustainable development: II. Corporate social responsibility and "stakeholder management" versus a rights-based approach to sustainable development, *Sustainable Development*, 9, pp. 121–135.

Boserup, E. (1961), *Population and Technological Change: A Study of Long-Term Trends*, Chicago, IL, University of Chicago Press.

Bozorgmehr, N. (2014), Iran: dried out, *Financial Times (UK)*, August 21.

Briscoe, J. (2011), Water as an economic good: old and new concepts and implications for analysis and implementation, In *Treatise on Water Science*, Wilderer, P. (ed.), Oxford, Elsevier, vol. 1.

Brooks, D. B. (2006), An operational definition of water demand management, *International Journal of Water Resources Development*, 22, pp. 521–528.

Bryant, N. (2008), Australia's food bowl lies empty, *BBC News*, March 11.

Buzan, B. (1997), Rethinking security after the Cold War, *Cooperation and Conflict*, 32, pp. 5–28.

Cannon, T. (2008), Vulnerability, ''innocent'' disasters, and the imperative of cultural understanding, *Disaster Prevention Management*, 17, pp. 350–357.

Cannon, T. and D. Muller-Mahn (2010), Vulnerability, resilience and development discourses in context of climate change, *Natural Hazards*, 55, pp. 621–635.

Cardozo, F. A. C. (2008), Teargas, batons break up demonstrations, *Arab Times (Kuwait)*, July 28.

Cassidy, P. (2004), Kuwait oil refining halted by blackout, *CBS MarketWatch*, October 31.

Charabi, Y. (2013). Projection of future changes in rainfall and temperature patterns in Oman. *Journal of Earth Science and Climate Change*, 4, p. 154.

Cha, V. D. (2000), Globalization and the study of international security, *Journal of Peace Research*, 37, pp. 391–403.

Chandra Das, K. and Nilambari, G. (2009). Localization of labor and international migration: a case study of the sultanate of Oman. International Institute for Population Sciences (IIPS), see http://iussp2009.princeton.edu/abstractViewer.aspx?submissionId= 91995,.

Chaoul, H. J. (2013), Saudi Arabia: the way ahead, Jeddah, Saudi Arabia, AlKhabeer Capital, November 18, see http://www.alkhabeer.com/reports/saudi-arabia-way-ahead.

Chatham House (2011), British attitudes towards the UK's international priorities, London, Chatham House–YouGov Survey July 2011, see http://www.chathamhouse.org/sites/default/files/0711ch_yougov_survey.pdf.

Chatila, F. (2011), Emergency water storage, *Arab Water World*, xxxv, p. 4.

Choucri, N. and R. North (1975), *Nations in Conflict: National Growth and International Violence*, San Francisco, CA, W. H. Freeman.

Choucri, N. and R. C. North (1989), Lateral pressure in international relations: concept and theory, in Midlarsky, M.I. (ed.), *Handbook of War Studies*, Ann Arbor, MI, University of Michigan Press.

Christy, R., E. Mabaya, N. Wilson, E. Mutambatsere, and N. Mhlanga (2009), Enabling environments for competitive agro-industries, in da Silva, C. A., D. Baker, A. W. Shepherd, C. Jenane, and S. Miranda-da-Cruz (eds.), *Agro-industries for Development*, Wallingford and Cambridge, MA, The Food and Agriculture Organization of the United Nations and The United Nations Industrial Development Organization, pp. 136–222.

CIA (2011). *CIA Fact Book*. Washington, DC, Central Intelligence Agency.

Cirincione, J. (2009), Chain reaction, *Foreign Policy*, May 7.

Coker, M. (2013), Gas-rich Qatar grapples with traffic tailbacks, *Wall Street Journal*, July 17.

Connell, M. (2010), Iran's military doctrine, in the Iran primer: power, politics, and US policy, United States Institute of Peace Press, see http://iranprimer.usip.org/resource/irans-military-doctrine.

Cordesman, A. H. (1997), *Kuwait: Recovery and Security After the Gulf War*, Boulder, CO, Westview Press.

Cordesman, A. H. (2011), *Saudi Stability in a Time of Change*, Washington, DC, Center for Strategic and International Studies.

Cowell, A. (1989), 5 are killed in south Jordan as rioting over food prices spreads, *New York Times*, April 20.

Crary, D. D. (1951), Recent agricultural developments in Saudi Arabia, *Geographical Review*, 41, pp. 366–383.

Crystal, J. (1995), *Oil and Politics in the Gulf: Rulers and Merchants in Kuwait and Qatar*, Cambridge, Cambridge University Press.

CSIS (2010), *Cultivating Global Food Security: A Strategy for US Leadership on Productivity, Agricultural Research, and Trade*, Washington, DC, Center for Strategic and International Studies.

Cummins, C. (2008), Price protests erupt in Mideast, *Wall Street Journal*, March 20, p. A12.

Daehler, C. C. and S. K. Majumdar (1992), Environmental impacts of the Persian Gulf War, in Majumdar, S.K., G.S. Forbes, E.W. Miller, and R.F. Schmalz (eds.), *Natural and Technological Disasters: Causes, Effects, and Preventive Measures*, Easton, PA, The Pennsylvania Academy of Science.

Davidson, C. (2012), The importance of the unwritten social contract among the Arab monarchies, *New York Times*, August 29.

Davidson, C. M. (2013). *After the Sheikhs: the coming collapse of the Gulf monarchies*. Oxford: Oxford University Press.

Davis, M. (2006), Fear and money in Dubai, *New Left Review*, 41, September–October, pp. 47–68.

den Hond, F. and F. G. A. de Bakker (2012), Boomerang politics: how transnational stakeholders impact multinational corporations in the context of globalization, in Lindgreen, A., P. Kotler, J. Vanhamme, and F. Maon (eds.), *A Stakeholder Approach to Corporate Social Responsibility: Pressures, Conflicts, Reconciliation*, Aldershot, Gower, pp. 275–292.

De Schutter, O. (2011), How not to think of land-grabbing: three critiques of large-scale investments in farmland, *The Journal of Peasant Studies*, 38, March, pp. 249–279.

Detrie, M. (2011), Indigenous grasses to reduce water waste, *The National (UAE)*, January 29.

Dilanian, K. (2011), Virtual war a real threat, *Los Angeles Times*, March 28.

D'Odorico P., F. Laio, and L. Ridolfi (2010), Does globalization of water reduce societal resilience to drought?, *Geophysical Research Letters*, 37, L13403, pp. 1–5.

Donald, B., M. Gertler, M. Gray, and L. Lobao (2010), Re-regionalizing the food system? *Cambridge Journal of Regions, Economy and Society*, 3, pp. 171–175.

Dorsey, J. M. (2010), US, Europe press GCC states on Yemen membership, *Middle East Studies*, October 13.

Dorsey, J. M. (2013), Qatar broaches sensitive demography through soccer – analysis, *Eurasia View*, March 19.

Drummond, J. (2011), Seed bank hunts for solutions to saltier soil, *Financial Times (UK)*, May 2.

Durodie, W. (2004), Facing the possibility of bioterrorism, *Current Opinion in Biotechnology*, 15, pp. 264–269.

Dziuba, M. (2011), *Scarcity and Strategy in the GCC*. Washington, DC, Center for Strategic and International Studies.

EAD (n. d.), Instructions for Farmers on Water Saving Irrigation Techniques, Environment Agency – Abu Dhabi, see http://www.ead.ae/en/elibrary/.

EAD (2009), *Abu Dhabi Water Master Plan*, Abu Dhabi, United Arab Emirates, Environment Agency of Abu Dhabi.

Economist (2000), Sudan's oil: fuelling a fire, *Economist*, August 31.

Economist (2001), The spider in the web [A trickier enemy is hard to imagine], *Economist*, September 20.

Economist (2008a), How the other fifth lives, *Economist*, April 3.

Economist (2008b), Buying the farm: feeding its own people more cheaply, *Economist*, August 21,

Economist (2009), Outsourcing's third wave, *Economist*, May 21.

Economist (2012), Let them eat baklava, *Economist*, March 17.

EIA (n.d. a), What drives crude oil prices? US Energy Information Administration, see http://www.eia.gov/finance/markets/supply-opec.cfm.

EIA (n.d. b), Bahrain, US Energy Information Administration, see http://www.eia.gov/countries/country-data.cfm?fips=ba#pet.

EIU (2010), The GCC in 2020: resources for the future, The Economist Intelligence Unit, see http://graphics.eiu.com/upload/eb/GCC_in_2020_Resources_WEB.pdf.

El Shammaa, D. (2008), Employers must give health insurance, *Gulf News (UAE)*, January 1.

El-tablawy, T. (2011), GCC pledges $20 billion in aid for Oman, Bahrain, *Washington Post*, March 10.

Embassy of Saudi Arabia (n.d.), Agricultural achievements, Washington, DC, see http://www.saudiembassy.net/about/country-information/agriculture_water/Agricultural_Achievements.aspx.

Emirates247 (2012), Illegal migrant arrested in Saudi for 41st time. *Emirates277*, April 23.

Emirates247 (2013), Expats born in Saudi must be given citizenship: scholar, *Emirates277*, November 1.

Emirates247 (2013b), Why UAE residents' water and electricity bill may surge soon, *Emirates277*, November 14.

EPA (2014), *Water-Saving Technologies, Water Sense*, Washington, DC, US Environmental Protection Agency, Office of Wastewater Management.

Estidama (n.d.), Government of the United Arab Emirates, see http://estidama.org/estidama-and-pearl-rating-system.aspx?lang=en-US

Evans, A. (2011), Brits think resource scarcity is a bigger deal than climate or development – survey, *Global Dashboard*, July 25.

Falkenmark, M. J., J. Rockström, and H. Savenjie (2004), *Balancing Water for Humans and Nature*, London, Earthscan.

FAO (1997), *Irrigation in the Near East Region in Figures*, Rome, Food and Agriculture Organization of the United Nations.

FAO (2002), *The State of Food Insecurity in the World 2001*. Rome, Food and Agriculture Organization of the United Nations.

FAO (2008), *Water Report 34, 2009, Water Profiles of Different GCC Countries*, Rome, Food and Agriculture Organization of the United Nations.

FAO (2011a), *The State of the World's Land and Water Resources for Food and Agriculture (SOLAW) – Managing Systems at Risk*, London, Earthscan and Rome, Food and Agriculture Organization of the United.

FAO (2011b), *The State of Food Insecurity in the World*, Rome, Food and Agriculture Organization of the United Nations.

FAO (2012), *Voluntary Guidelines on the Responsible Governance of Tenure of Land, Fisheries and Forests in the Context of National Food Security*, Rome, Food and Agriculture Organization of the United Nations.

FAOstat (n.d.) Statistics Division of FAO, Food and Agriculture Organization, see FAO-STAT.fao.org.

Faruqi, N. I. (2001), Intersectoral water markets in the Middle East and North Africa, in, *Water Management in Islam*, Faruqui, N. I., A. K. Biswas, and M. J. Bino (eds.), New York, NY, United Nations University Press.

Faruqui, N. I. (2003), Water, human rights, and economic instruments: the Islamic perspective, *Water Nepal*, 9/10, pp. 197–214.

Fasano, U. (2002), With open economy and sound policies, UAE has turned oil "curse" into a blessing, *IMF Survey*, 31, October 21, pp. 330–332.

Fasano, U. and Z. Iqbal, (2003), GCC countries: from oil dependence to diversification, *International Monetary Fund*, see http://www.imf.org/external/pubs/ft/med/2003/eng/fasano/.

Fattouh, B. and L. Mahadeva (2014), *Price Reform in Kuwait's Electricity and Water: Assessing the Net Benefits in the Presence of Congestion*, Oxford, Oxford Institute for Energy Studies.

Fayad, W., N. Batri, and J. Ayoub (2011), Gulf must act to stem heavy water use, *Financial Times (UK)*, October 5.

Feagan, R. (2007), The place of food: mapping out the "local" in local food systems, *Progress in Human Geography*, 31, pp. 23–42.

Fischer, G., H. van Velthuizen, M. Shah, and F. Nachtergaele, (2002), *Global Agro-ecological Assessment for Agriculture in the 21st Century: Methodology and Results*, Kaxenburg, Austria and Rome, International Institute for Applied Systems Analysis and Food and Agriculture Organization of the United Nations.

Fisher, I. (1999), Oil flowing in Sudan, raising the stakes in its civil war, *New York Times*, October 17.

Flintoff, C. (2014), Russia aims to implement the tightest security in Olympic history, *National Public Radio (USA)*, January 15.

Forcese, C. (2001), Militarized commerce, *Canadian Foreign Policy*, Spring, 8, pp. 37–56.

Forstenlechner, I. and E. Rutledge (2010), Unemployment in the Gulf: time to update the "social contract", *Middle East Policy*, 17, pp. 38–51.

Forsythe, J. (2011), *Middle East and North Africa Program paper MENA, Opportunities and Obstacles for Yemeni Workers in GCC Labor Markets*, London, Chatham House.

Forsythe, M. (2012), China workers abroad becoming easy prey, *Bloomberg News*, February 1.

Freedom House (2014), *Freedom in the World 2014*, Washington DC, Freedomhouse.org.

Friedman, L. (2011), Middle East's push toward renewable energy spurred by rising oil prices, *New York Times*, June 21.

Fritz, H. M, C. D. Blount, F. B. Albusaidi, and A. H. M. Al-Harthy (2010). Cyclone Gonu storm surge in Oman, *Estuarine, Coastal and Shelf Science*, 86, pp. 102–106.

Fuchs, M. (2012), Qatar's next big purchase: a farming sector, *Reuters*, January 6.

Funk, M. (2010), Will global warming, overpopulation, floods, droughts and food riots make this man rich? Meet the new capitalists of chaos, *Rolling Stones*, May 27.

Furedi, F. (2007), The changing meaning of disaster, *Area*, 39, pp. 482–489.

Gallopin, G. (2006), Linkages between vulnerability, resilience and adaptive capacity, *Global Environmental Change*, 16, pp. 293–303.

Gause, F. G., III (2011), Why Middle East studies missed the Arab Spring: the myth of authoritarian stability, *Foreign Affairs*, 90, pp. 81–90.

Ghafour, P. K. A. (2010), Solar energy initiative launched, *Arab News*, January 25.

Giddens, A. (1991), *Modernity and Self-Identity: Self and Society in the Late Modern Age*, Cambridge, Polity.

Gilmore, C. (2012a), Filtration system is high-tech, *My Daily News (Australia)*, February 4.

Gilmore, C. (2012b), Water bans as supply system fails, *My Daily News (Australia)*, January 27.

Gladman, A. (1997), Massive Nile River diversions planned, *World Revivers International*, 12, June.

Gleditsch, N. P. (2001), Armed conflict and the environment, in Diehl, P.F. and N.P. Gleditsch (eds.), *Environmental Conflict*, Boulder, CO, Westview.

Gleick, P. (n.d.), Pacific Institute, Water Conflict Chronology. California, CA, see http://www2.worldwater.org/chronology.html.

Gleick, P. H. (2009). The question is irrelevant, *Seed Magazine*, May 14.

Gleick, P. H., P. Yolles, and H. Hatami (1994), Water, war and peace in the Middle East, *Environment*, 36, pp. 6–15.

Global Water Partnership (2000), *Towards Water Security: A Framework for Action*, Stockholm, Global Water Partnership (GWP).

Goldstone, J. A. (2011), Cross-class coalitions and the making of the Arab revolts of 2011, *Swiss Political Science Review*, 17, pp. 457–462.

Government of Abu Dhabi (2008), *The Abu Dhabi Economic Vision 2030*, Abu Dhabi.

Government of Bahrain (2008), *From Regional Pioneer to Global Contender: The Economic Vision 2030 for Bahrain*, Kingdom of Bahrain.

GRAIN (2008), Seized! GRAIN briefing annex, GRAIN Briefing, October. See http://www.grain.org/media.

Grey, D. and Sadoff, C. (2007), Sink or swim? Water security for growth and development. *Water Policy*, 9, pp. 545–571.

Grey, D, D. Garrick, D. Blackmore, *et al.* (2013), Water security in one blue planet: twenty-first century policy challenges for science. *Philosophical Transactions of the Royal Society A*, 371, pp.1–22.

Gulf Daily News (2008), Bahrain to buy 'farms' in Saudi, *Gulf Daily News*, February 4.

Gulf News (2012), Farms stop cultivating Rhodes grass, *Gulf News (UAE)*, March 23.

Gurman, H. (2011), Migrant workers in Libya. *Independent World Report*, April 25.

Gutub, S. A., M. F. Soliman, and A. Uz Zaman (2013). Saudi Arabia confronts with water scarcity: an insight, *International Journal of Water Resources and Arid Environments*, 2, pp. 218–225.

GWI (2003), Reducing unaccounted-for water, *Global Water Intelligence*, 4, Issue 4 (April).

Habboush, M. (2010), Drumming up support for national identity, *The National (UAE)*, February 21.

Halliday, F. (1978), Decline in Oman: PFLO's new political strategy, *MERIP Reports*, No. 67, May, pp. 18–21.

Hamdan, S. (2013), United Arab Emirates looks to vocational education, *New York Times*, November 24.

Hauslohner, A. (2011), Among Libya's prisoners: interviews with mercenaries, *Time (magazine)*, February 22.

Headey, D. and S. Fan (2010), Reflections on the global food crisis, how did it happen? How has it hurt? And how can we prevent the next one? *International Food Policy Research Institute, Research Monograph*, 165.

Hedden, W. P. (1929), *How Great Cities are Fed*, Boston, MA, D.C. Heath and Company.

Heymann, E. (2010), *World Water Markets*, Frankfurt, Deutsche Bank Research, June 1.

Himes, J. C. (2011), *Iran's Maritime Evolution*, Washington, DC, Center for Strategic and International Studies.

Hoekstra, A. and A. Chapagain (2008), *Globalization of Water: Sharing the Planet's Freshwater Resources*, Malden, MA, Wiley-Blackwell.

Hoff, H. (2011), *Understanding the Nexus*. Background Paper for the Bonn 2011 Conference: The Water, Energy and Food Security Nexus, Stockholm, Stockholm Environment Institute.

Homer-Dixon, T. (1999), *Environment, Scarcity, and Violence*, Princeton, NJ: Princeton University Press.

Hope, B. (2011), Egypt freezes Kingdom farm land deal, *The National (UAE)*, April 12.

Horne, F. (2011), *Understanding Land Deals in Africa, Country Report: Ethiopia*, Oakland, CA, The Oakland Institute.

Hubbard, G. P. and A. P. O'Brien (2012), *Microeconomics*, 4th Edition, Upper Saddle River, NJ, Prentice Hall.

Hurlimann, A. (2006). Water, water, everywhere – which drop should be drunk? *Urban Policy and Research*, 24, pp. 303–305.

Husain, A. and K. Habib (2005), Investigation of tubing failure of super-heater boiler from Kuwait desalination electrical plant, *Desalination*, 183, pp. 203–208.

ICA (2012) *Global Water Security*, Intelligence Community Assessment, ICA 2012–08, February 2, see https://www.fas.org/irp/nic/water.pdf.

ICG (2003), *Yemen: Coping With Terrorism and Violence in a Fragile State*, Amman and Brussels, International Crisis Group (ICG), ICG Middle East Report N°8, January 8.

ILO (2011), *Global Employment Trends 2011: The Challenge of a Jobs Recovery*, Geneva, International Labor Office.

IMF (2012a), *Economic Prospects and Policy Challenges for the GCC Countries*, Gulf Cooperation Council: Annual Meeting of Ministers of Finance and Central Bank Governors, October 5–6, Saudi Arabia, Washington, DC, International Monetary Fund.

IMF (2012b), *Global Monitoring Report 2012, Food Prices, Nutrition, and the Millennium Development Goals*, Washington, DC, International Monetary Fund and World Bank.

IMF (2013), *Saudi Arabia: Selected Issues, IMF Country Report No. 13/230, July*, Washington, DC, International Monetary Fund and World Bank.

Immerzeel, W., P. Droogers, W., Terink, et al. (2011), *Middle-East and Northern Africa Water Outlook, April, FutureWater Report: 98*, Wageningen, FutureWater.

IRENA (2012), *Water Desalination Using Renewable Energy: Technology Brief*, The International Renewable Energy Agency (IRENA), Technology Brief I12.

Jagannathan, N. V., A. S. Mohamed, and A. Kremer, (eds.) (2009), *Water in the Arab World: Management Perspectives and Innovations*, Washington, DC, The International Bank of Reconstruction and Development, The World Bank.

Jackson, R. B., S. R. Carpenter, C. N. Dahm, *et al.* (2001), Water in a changing world, *Ecological Applications*, 11, pp. 1027–1045.

Jägerskog, A., A. Swain, and J. Öjendal (2014), *Water Security – International Conflict and Cooperation*, London, Sage Publications.

Jones, T. C. (2010), *Desert Kingdom: How Oil and Water Forged Modern Saudi Arabia*, Cambridge, MA, Harvard University Press.

Jury, W. A. and Vaux, H. Jr. (2005), The role of science in solving the world's emerging water problems. *Proceedings of the National Academy of Sciences (PNAS)*, 102, pp. 15715–15720.

Kadhim, A. (2006), United we stand, *AlAhram Weekly*, April 13–19, Issue No. 790.

Kaikati, J. G. (1980), The economy of Sudan: a potential breadbasket of the Arab world? *International Journal of Middle East Studies*, 11, pp. 99–123.

Kamrava, M. (2012), The Arab Spring and the Saudi-led counterrevolution, *Orbis*, Winter, pp. 96–104.

Kanady, S. (2013), Food security plan getting final touches, *The Peninsula (Qatar)*, June 30.

Kassem, M. S. (1989), Strategy formulation: Arabian Gulf style, *International Studies of Management and Organization*, 19, pp. 6–21.

Katzman, K. (2010), The United Arab Emirates (UAE): issues for US policy, *Congressional Research Service*, June 23.

Katzman, K. (2011), Bahrain: reform, security, and US policy, *Congressional Research Service*, December 29.

Keck, M. E. and K. Sikkink, (1998), *Activists Beyond Borders: Advocacy Networks in International Politics*, Ithaca, NY, Cornell University Press:

Kerr, S. (2012), Agricel aims to make desert yield crops, *Financial Times (UK)*, March 19.

Kerr, S. and S. Mishkin (2011), "Expat fat" cut as Abu Dhabi shifts focus, *Financial Times (UK)*, June 27.

Khaleej Times (2011), UAE moves towards building stronger union, *Khaleej Times (UAE)*, December 1.

Khaleej Times (2012), Law to rationalise water, energy use drafted, *Khaleej Times (UAE)*, September 27.

Khaleej Times (2012b), 2013: year for Emiratisation, *Khaleej Times (UAE)*, November 27.

Kloppenburg, J., J. Hendrickson, and G.W. Stevenson (1996), Coming in to the foodshed, in Vitek, W. and W. Jackson (eds.), *Rooted in the Land: Essays on Community and Place*. New Haven, CT, Yale University Press, pp. 113–123.

Knowledge@Wharton (2011), *To Stave Off Arab Spring Revolts, Saudi Arabia and Fellow Gulf Countries Spend $150 Billion*, The Wharton School, The University of Pennsylvania, September 21.

Kotilaine, J. T. (2010), *GCC Agriculture*, Riyadh, Kingdom of Saudi Arabia, National Commercial Bank Capitals, March.

Krane, J. (2012), The end of the Saudi oil reserve margin, *Wall Street Journal*, April 3.

Kumar, A. (2009), Reclaimed islands and new offshore townships in the Arabian Gulf: potential natural hazards, *Current Science*, 96, pp. 480–485.

Kuttner, R. (1987), *The Economic Illusion: False Choices Between Prosperity and Social Justice*, Philadelphia, PN, University of Pennsylvania Press.

Kumetat, D. (2012), Climate change on the Arabian Peninsula – regional security, sustainability strategies, and research needs, in, J. Scheffran *et al.* (eds.), *Climate Change, Human Security and Violent Conflict, Hexagon Series on Human and Environmental Security and Peace 8*, Berlin, Springer-Verlag.

Kutbi, I. I. and K. M. Al Sulaiman (1994), Features of nuclear desalination systems for Saudi Arabia, *Desalination*, 97, pp. 327–337.

Kuwait Times (2008), Expats numbers to be restricted, *Kuwait Times*, August 7.

Kuznets, S. (1966), *Modern Economic Growth: Rate, Structure, and Spread*, New Haven, CT, Yale University Press.

Kyriakakis, J. (2007), Australian prosecution of corporations for international crimes: the potential of the Commonwealth criminal code, *Journal of International Criminal Justice*, 5, pp. 809–826

Lacroix, S. (2011), Is Saudi Arabia immune? *Journal of Democracy*, 22, 48–59.

Lahn, G. and P. Stevens (2011), *Burning Oil to Keep Cool: The Hidden Energy Crisis in Saudi Arabia*, Chatham House, London, The Royal Institute of International Affairs.

Lahn, G., P. Stevens, and F. Preston (2013), *Saving Oil and Gas in the Gulf*, Chatham House, London, The Royal Institute of International Affairs.

Landais, E. (2007), Water "is running out because it is being provided for free", *Gulf News (UAE)*, February 6.

Landais, E. (2008), Dubai introduces new rates to curb use of electricity and water, *Gulf News (UAE)*, February 17.

Langelaan, H. C., F. Pereira da Silva, U. Thoden van Velzen, *et al.* (2013), Technology options for feeding 10 billion people: options for sustainable food processing, State of the art report, November, see http://www.europarl.europa.eu/RegData/etudes/etudes/join/2013/513533/IPOL-JOIN_ET(2013)513533_EN.pdf.

Levins, J. M. (1995), The Kuwaiti Resistance, *Middle East Quarterly*, March 2, pp. 25–36.

Levy, G. (2006), As the Hamas team laughs, *Haaretz (Israel)*, February 19.

Li, T. M. (2011), Centering labor in the land grab debate, *Journal of Peasant Studies*, 38, Issue 2.

Lidstone, D. (2010), Gulf buries its head in the sand over water, *Financial Times (UK)*, January 18.

Lipscombe, J. (2014), Fallout from the Saudi Aramco breach continues, *Bloomberg Business Week, Middle East*, August 4.

Livescience.com (2012), Fossil footprints reveal oldest elephant herd, *Livescience*, February 22.

Liu, B., X. Mei, Yu Li, and Y. Yang (2007), The connotation and extension of agricultural water resources security. *Agricultural Sciences in China*, 6, pp. 11–16.

Lloyd, J. W., J. G. Pike, B. L. Eccleston, and T. R. E. Chidley (1987), The hydrology of complex lens conditions in Qatar, *Journal of Hydrology*, 89, pp. 239–258.

Lori, N. (2012), *Temporary Workers or Permanent Migrants? The Kafala System and Contestations over Residency in the Arab Gulf States*, Paris, Institut français des relations internationales (Ifri), Center for Migrations and Citizenship.

Lowe, A. (2011), Investing abroad to secure food at home. *Gulf News*, March 9.

Lynch, M. (2011), Will the GCC stay on top? *Foreign Policy*, December 15.

MacDonald, F. (2010), Kuwait's power consumption hits record as temperature reaches 50 Celsius, *Bloomberg*, June 13.

Macdonald, K. (2009), *The Reality of Rights: Barriers to Accessing Remedies When Business Operates Beyond Borders*, London, The London School of Economics and Political Science (LSE).

Mackinder, H. J. (1904), The geographical pivot of history, *Geographical Journal*, 23, pp. 161–169.

MAF and ICBA (2012), *Oman Salinity Strategy, Agricultural Status and Salinity Impact*, Ministry of Agriculture and Fisheries (Oman), and International Center For Biosaline

Agriculture (UAE), see maf.gov.om/Download.ashx?File=FCKupload/File/books/annex2.pdf.

Mail and Guardian (2007), Omani capital's water restored following cyclone, *Mail and Guardian (Johannesburg)*, June 12.

Malik, A. U., P. C. M. Kutty, I.N. Andijani, and S.A. Al-Fozan (1994), Materials performance and failure evaluation in SWCC MSF plants, *Desalination*, 97, p. 171.

Maree, K. (2008), Riyadh targets $11bn in water savings, *Middle East Economic Digest*, March 20.

Mathews, R. (2005), A six-step framework for ecologically sustainable water management, *Universities Council on Water Resources Journal of Contemporary Water Research and Education*, pp. 60–65, June.

Matthews, D. (2013), We've tried guest worker programs before. They don't work, *Washington Post*, January 30.

Matthiesen, T. (2013), EU foreign policy towards Bahrain in the aftermath of the uprising, in Echagüe, A. (ed.), *The Gulf States and the Arab Uprisings, FRIDE and the Gulf Research Center*, Madrid, Artes Gráficas Villena.

Mazzetti, M. and E. B. Hager (2011), Secret desert force set up by Blackwater's founder, *New York Times*, May 14.

McPhail, A., A. R. Locussol, and C. Perry (2012), *Achieving Financial Sustainability and Recovering Costs in Bank Financed Water Supply and Sanitation and Irrigation Projects*, Washington, DC, World Bank.

Mekonnen, D.Z. (2010), The Nile Basin cooperative framework agreement negotiations and the adoption of a "water security" paradigm: flight into obscurity or a logical cul-de-sac? *European Journal of International Law*, 21, pp. 421–440.

Meenaghan, G. (2013), Qatar to act on expat workers' rights, *The National (UAE)*, November 18.

Metz, H. A. (1993), *Persian Gulf States: A Country Study*, Washington, GPO for the Library of Congress.

Millington, B. (2009), 7,000 Bahrain workers strike "in pay dispute," *Arabian Business*, June 10.

Micklethwait, J. and A. Wooldridge (2014), The state of the state: the global contest for the future of government, *Foreign Affairs*, July–August, 93, pp. 118–132.

Milutinovic, M. (2006), Water demand management in Kuwait, Master's Thesis, Cambridge, MA, Massachusetts Institute of Technology, Department of Civil and Environmental Engineering.

Momtaz, R. (2012), Officials: Iran, Syria aided bomb, assassination plot in Bahrain, *ABC News*, January 31.

Moore, P. W. and B. F. Salloukh (2007), Struggles under authoritarianism: regimes, states, and professional organizations in the Arab world, *International Journal of Middle East Studies*, 39, pp. 53–76.

Mousseau, F. (2010), *The High Food Price Challenge: A Review of Responses to Combat Hunger*, Oakland, CA, The Oakland Institute.

Murad, A. A., H. Al Nuaimi, and M. Al Hammadi (2007), Comprehensive assessment of water resources in the United Arab Emirates (UAE). *Water Resources Management*, 21, pp. 1449–1463.

Murakami, M. (1995), *Managing Water for Peace in the Middle East: Alternative Strategies*, Tokyo and New York, United Nations University Press.

Musayab, R. (1988), *Water Resources and Development in the State of Bahrain*. Directorate of Water Supply, Ministry of Electricity and Water.

Nair, P. (2013), Ablution solutions, *BuildGreen*, March 7, pp. 38–40.

National (2014), Re-using waste water must be a UAE priority, *National (UAE)*, January 16.

Netting, R. (1993), *Smallholders, Householders: Farm families and the Ecology of Intensive, Sustainable Agriculture*, Stanford, CA, Stanford University Press.

Niasse, M. and Taylor, M. (2010). Building an informed and inclusive response to the global rush for land, World Bank Annual Conference on Land Policy and Administration Washington, DC, April 26–27.

Niblock, T. (2007), *The Political Economy of Saudi Arabia*, London, Routledge.

Nkrumah (2011), Libya's blacks on the run, *Al-Ahram (Cairo)*, September 8–14.

North, R. (1984), Integrating the perspectives: from population to conflict and war, in Choucri, N. (ed.), *Multidisciplinary Perspectives on Population and Conflict*, Syracuse, NY, Syracuse University Press, pp. 195–215.

Oakland Institute (2014), *Walking on the West Side: The World Bank and the IMF in the Ukraine Conflict*, Oakland, CA, The Oakland Institute.

OBG (2008), Saudi Arabia: national water, *The Middle East and North Africa Financial Network*, February 10.

OBG (2013), Saudi Arabia: overseas investments to boost food security, *Oxford Business Group*, May 9.

O'Brien, E. M. (2006). Biological relativity to water-energy dynamics. *Journal of Biogeography*, 33, pp. 1868–1888

OOSKAnews (2013), Saudi Arabia ups desal production by 2 million cubic meters, *OOSKAnews*, April 5.

Orr, S., A. Cartwright, and D. Tickner (2009), *Understanding Water Risks: a Primer on the Consequences of Water Scarcity for Government and Business*, Godalming, WWF-UK.

Oxfam (2011), Land and power: the growing scandal surrounding the new wave of investments in land, *Oxfam International*, September 22.

Paarlberg, R. L. (1985), *Food Trade and Foreign Policy: India, the Soviet Union, and the United State*, Ithaca, NY, Cornell University Press.

Parker, R. S. (2010), *Water and Development: An evaluation of World Bank Support, 1997–2007, Vol. 1 of Water and Development: An Evaluation of World Bank Support, 1997–2007, IEG Study Series, No. 1.*, Washington, DC, World Bank.

Parris, K. (2010), *Sustainable Management of Water Resources in Agriculture*, Paris, Organization for Economic Co-operation and Development (OECD).

Patel, R. (2007), *Stuffed and Starved: The Hidden Battle for the World Food System*, London, Portobello Books Ltd.

Pearce, F. (2012), *Land Grabbers: The New fight Over who Owns the Earth*, Boston, MA: Beacon Press.

Pearce, F. (2009). A common source of grievance, *Seed Magazine*, May 14.

Peninsula (2011a), Qatar plans 70pc self-sufficiency in food by 2023, *Peninsula (Qatar)*, July 28.

Peninsula (2011b), Reform only way to avoid revolts: PM, *Peninsula (Qatar)*, November 10

Perez, I. and ClimateWire (2013), Climate change and rising food prices heightened Arab Spring, *Scientific American*, March 4.

Perlroth, N. (2012), *In cyberattack on Saudi Firm, US sees Iran firing back, New York Times*, October 23.

Perrow, C. (1984), *Normal Accidents: Living with High-Risk Technologies*, New York, NY, Basic Books.

Perrow C. (1999), *Normal Accidents with an Afterword and Postscript on Y2K*, Princeton, NJ, Princeton University Press.

Perrow, C. (2011), Fukushima and the inevitability of accidents, *Bulletin of the Atomic Scientists*, 67, pp. 44–52.

Peters, C. J., N. L. Bills, J. L. Wilkins, and G. W. Fick (2009), Foodshed analysis and its relevance to sustainability, *Renewable Agriculture and Food Systems*, 24, pp. 1–7.

Postel, S. (2009). But who will export tomorrow's virtual water?, *Seed Magazine*, May 14.

Priscoli, J. D. (2012), Introduction, *Water Policy*, 14, pp. 3–8.

QGSDP (2011), *Qatar National Development Strategy 2011–2016: Towards Qatar National Vision 2030*, Doha, Qatar, Qatar General Secretariat for Development Planning (QGSDP).

Quist-Arcton, O. (2011), In Libya, African migrants say they face hostility, *National Public Radio News*, February 25.

Radan, S. (2013), New guide to boost Emirati–expat ties, *Khaleej Times*, January 3.

Randeree, K. (2012), Workforce nationalization in the Gulf Cooperation Council States, *Center for International and Regional Studies, Georgetown University School of Foreign Service in Qatar, Occasional Paper*, No. 9.

Relph, E. C. (1976), *Place and Placelessness*, London, Pion.

Renaud, F. and L. Wirkus, (2012), Water, climate change and human security: conflict and migration, In *The Global Water Crisis: Addressing an Urgent Security Issue*, Bigas, H. (ed.), Hamilton, Ontario: UNU-INWEH.

Renewables International (2014), German power sector 27 percent non-hydro renewable in 2014, *Renewables International*, July 3.

Reuters (2007), Iran won't use oil as weapon if attacked: Ahmadinejad, *Reuters (US edition)*, November 18.

Reuters (2010) Abu Dhabi fears terror threat to water supply, *Arabian Business*, December 20.

Reuters (2011a), Qatar hikes salaries, pensions for state employees, *Arabian Business*, September 7.

Reuters (2011b), Middle East unrest may scare off expat workers, *Arabian Business*, February 9.

Reuters (2011c), Gulf political, economic reform crucial: Kuwait PM, *The Daily Star (Lebanon)*, July 5.

Reuters (2012). Gulf states must tackle Muslim Brotherhood threat: UAE. *Reuters*, Oct 8.

Reuters (2013a), GCC seeks food security in Europe after African problems, *TradeArabia*, December 30.

Reuters (2013b), Kuwait to review $16bn annual subsidies, *Gulf Business*, November 11.

Reuters (2013c), Gulf Arab states eye Arabian Sea for safer water supplies, *Gulf Business*, July 18.

Reuters (2014a), Call to 'rationalize' subsidy programs, *Arab News (Saudi Arabia)*, April 22.

Reuters (2014b), Saudi Arabia needs to spend $213bn on power, water projects, *Arabian Business*, May 13.

Reuters (2014c), UAE–Egypt alliance expands to desert wheat venture, *Reuters*, December 5.

Richter, B. D. (2010), Re-thinking environmental flows: from allocations and reserves to sustainability boundaries, *River Research and Applications*, 26, pp. 1052–1063.

Riedel, B. (2012), Iran seeks to exploit Shia grievances in Saudi Arabia, *Al-Monitor*, November 9.

Rodriguez, D. J., C. van den Berg, and A. McMahon (2012), *Investing in Water Infrastructure: Capital, Operations and Maintenance*, Washington, DC, World Bank.

Roderick, P. (2005) *A Human Rights, Environmental and Economic Monstrosity, Gas Flaring in Nigeria*, Amsterdam, Environmental Rights Action, and Climate Justice Program.

Rosenau, J. N. (2003), Globalization and governance: bleak prospects for sustainability, *International Politics and Society*, 3, pp. 11–29.

Rosenau, J. N. (2006), *The Study of World Politics, Volume 2: Globalization and Govern-ance*, New York, Routledge, Taylor, and Francis.

Ross, M. L. (2001), Does oil hinder democracy? *World Politics*, 53, pp. 325–361.

Rothschild, E. (1976), Food politics, *Foreign Affairs*, 54, pp. 285–307.

Roudi-Fahimi, F., L. Creel, and R. De Souza (2002), *Finding the Balance: Population and Water Scarcity in the Middle East and North Africa, July*, Washington, DC, Popula-tion Reference Bureau.

Ruckstuhl, S. (2009), *Renewable Natural Resources: Practical Lessons for Conflict-Sensitive Development*. Social Development Department, Sustainable Development Network, The World Bank Group.

Ruhl, B. S., V. Buckingham, and E. Pencak (2003), *Homeland Security and Drinking Water: An Opportunity for Comprehensive Protection of a Vital Natural Resource*, Washington, DC, Environmental Law Institute.

Ruhs, M. (2009), *Migrant Rights, Immigration Policy and Human Development*, United Nations Development Programme, Human Development Reports, Research Paper, 2009/23.

Russell, S. S. (1992), International migration and political turmoil in the Middle East, *Population and Development Review*, 18, pp. 719–727.

Russell, S. S. and M. A. Al-Ramadhan (1994), Kuwait's migration policy since the Gulf crisis, *International Journal of Middle East Studies*, 26, pp. 569–587.

Sadr, K. (2001), Water markets and pricing in Iran, in *Water Management in Islam*, Faruqui, N. I., A. K. Biswas, and M. J. Bino (eds), New York, NY, United Nations University Press.

Said, S. (2014), UAE oil minister: cut energy subsidies to lower domestic consumption, *Wall Street Journal*, January 21.

Saigol, L. (2011), Foreign companies face Arab Spring fallout, *Financial Times (UK)*, October 3.

Saleh, H. (2008), Invest oil money in food, UN says, *Financial Times (UK)*, March 6.

Salman, S. M. A. (2011), The new state of South Sudan and the hydro-politics of the Nile Basin, *Water International*, 36, pp. 154–166.

Saracini, N. (2011), *Stolen Land Stolen Future: Cambodia For Sale*, Brussels, APRODEV (Association of World Council of Churches related Development Organizations in Europe.

Saraf, A. (2013), Plug the leaks, *Utilities Middle East*, December 19.

Sathish, V. M. (2010), Demand for quality labor housing rises, *Emirates247*, May 10.

Saudi Gazette. (2014), SR300b for 8.5m m^3 desalinated water, *Saudi Gazette*, September 19.

Savenije, H. H. G. (2002), Why water is not an ordinary economic good, or why the girl is special, *Physics and Chemistry of the Earth*, 27, pp.741–744.

Schollaert, A. and D. van de Gaer (2009), Natural resources and internal conflict, *Environ-mental and Resource Economics*, 44, pp. 145–165.

Sdralevich, C., R. Sab, Y. Zouhar, and G. Albertin (2014) *Subsidy Reform in the Middle East and North Africa: Recent Progress and Challenges Ahead*, Washington, DC, International Monetary Fund.

Sen, A. (1981), *Poverty and Famines: An Essay on Entitlement and Deprivation*, Oxford, Clarendon Press.

Shah, N. M. (2006), Restrictive labor immigration policies in the oil-rich Gulf: effective-ness and implications for sending Asian countries, United Nations Expert Group Meeting on Social and Economic Implications of Changing Population Age Structure, Mexico City, August 31–September.

Shah, N. M. (2008), *Recent Labor Immigration Policies in the Oil-Rich Gulf: How Effective are they Likely to Be?* Ithaca, NY, Cornell University ILR School.

Shah, N. M. (2012), Socio-demographic transitions among nationals of GCC countries: implications for migration and labor force trends, *Migration and Development*, 1, pp. 138–148.

Shehab, F. (1964), Kuwait: a super-affluent society, *Foreign Affairs*, April. 3, pp. 461–474.

Shenaifi, M. S. (2013), Attitudes of students at College of Food and Agricultural Sciences toward agriculture, *Journal of the Saudi Society of Agricultural Sciences*, 12, pp. 117–120.

Shepherd, J. (1985), Ethiopia: the use of food as an instrument of US foreign policy, *Issue: A Journal of Opinion*, 14, pp. 4–9.

Sheppard, G. (1994), Disaster (hurricane) preparedness Public (water) utility, *Desalination*, 98, pp. 479–483.

Sherratta, A. (1980), Water, soil and seasonality in early cereal cultivation, *World Archaeology*, 11, pp. 313–330.

Shetty, S. (n. d.), Future vision of the Saudi economy, agriculture and water resources management: issues and options, World Bank, see https://uqu.edu.sa/files2/tiny_mce/plugins/filemanager/files/4150111/fourm/Shetty%20Report.pdf.

Siddiqi, A. and L. D. Anadon (2011), The water–energy nexus in Middle East and North Africa, *Energy Policy*, 39, August, pp. 4529–4540.

Smith, D. (2011), Shell accused of fuelling violence in Nigeria by paying rival militant gangs, *The Guardian*, October 2.

Solomon, S. (2010), *Water: The Epic Struggle for Wealth, Power and Civilization*, New York, NY, Harper Perennial.

Sophia, M. (2014), GCC mulls common water, power policy, *Gulf Business*, January 13.

Sorokin, P. A. (1957), *Social and Cultural Dynamics; a Study of Change in Major Systems of Art, Truth, Ethics, Law, and Social Relationships*, Boston, MA, Extending Horizons Books.

Sprout, H. (1963), Geopolitical hypotheses in technological perspective, *World Politics*, 15, pp. 187–212.

Sternberg, T. (2013), Chinese drought, wheat, and the Egyptian uprising: how a localized hazard became globalized, In Werrell, C. E. and F. Femia (eds.), *The Arab Spring and Climate Change: A Climate and Security Correlations Series*, Washington, DC, Center for American Progress.

Stevens Jr., G. P. (1949), Saudi Arabia's petroleum resources, *Economic Geography*, July 25, pp. 216–225.

Stratfor (2004), Iraq industry: new dangers to oil infrastructure? *Strategic Forecasting*. April 27.

Sullivan, J. K. (2011), *Water Sector Interdependencies, Summary Report*, Alexandria, VA, Water Environment Federation.

Surk, B. (2007), Dubai strike threatens building boom, *US Today*, October 28.

Swain, A. (2001), Water wars: fact or fiction, *Futurers (London)*, no. 33, October, pp. 769–781.

Swain, A. (2012), *Understanding Emerging Security Challenges: Threats and Opportunities*. London, Routledge.

Switzer, J. (2002), Oil and violence in Sudan, International Institute for Sustainable Development, and the IUCN-World Conservation Union Commission on Environmental, Economic and Social Policy, April 15, see https://www.iisd.org/pdf/2002/envsec_oil_violence.pdf

Tago, A. (2014), KSA water consumption rate twice the world average, *Arab News (Saudi Arabia)*, February 28.

Thompson, E. P. (1971), The moral economy of the English crowd in the eighteenth century, *Past and Present*, 50, pp. 76–136.

Thompson, R. (2008), Special report: power and water – controlling resources, *Middle East Economic Digest*, March 7.

Tindall, J. A. and A. A. Campbell (2010), Water security: national and global issues, *US Geological Survey, Fact Sheet*, 2010–3106, November, pp. 1–6.

Torbati, Y. (2013), Iran plans to build more nuclear reactors in quake-prone area, *Reuters*, April 10.

Toumi, H. (2013), Kuwait denies blacklisting Egyptians, *Gulf News (UAE)*, July 8.

Transparency International (2013), Corruption Perceptions Index 2013, see www.transparency.org.

Ulrichsen, K. C. (2009), Internal and external security in the Arab Gulf states, *Middle East Policy*, 16, pp. 39–58.

UN (2001), *Situation of Human Rights in the Sudan, Special Rapporteur of the Commission on Human Rights*, United Nations General Assembly, September 7, A/56/336.

UNDP (2004), *Water Governance for Poverty Reduction*, New York, NY, United Nations Development Programme.

UNDP (2009), *Arab Human Development Report*, New York, NY, United Nations Development Programme.

UNDP (2013). *Water Governance in the Arab Region: Managing Scarcity and Securing the Future*. New York, NY, United Nations Development Programme,

UNDP (2014), *Human Development Report 2014: Sustaining Human Progress: Reducing Vulnerabilities and Building Resilience*, New York, NY, United Nations Development Programme.

UNEP (n.d. a), *Sourcebook of Alternative Technologies for Freshwater Augmentation in West Asia*, United Nations Environmental Programme Newsletter and Technical Publications, see http://www.unep.or.jp/ietc/publications/techpublications/techpub-8f/b/cloud.asp.

UNEP (n.d. b), *Environmental Assessment of Ogoniland*, Nairobi, United Nations Environmental Programme, see http://postconflict.unep.ch/publications/OEA/UNEP_OEA.pdf.

USAID (2014). *USAID Water and Development Strategy 2013-2018*, United States Agency for International Development, see http://www.usaid.gov/sites/default/files/documents/1865/USAID_Water_Strategy_3.pdf.

US Department of State (2004), Attack on US Consulate General in Jeddah, US Department of State, On-the-Record Briefing, Jeddah, Saudi Arabia, December 7.

US Department of State (2011), *2010 Human Rights Report: Saudi Arabia*, US Department of State, Bureau of Democracy, Human Rights, and Labor, 2010 Country Reports on Human Rights Practices, April 8, see http://www.state.gov/j/drl/rls/hrrpt/2010/nea/154472.htm.

US Department of State (2012a), *Qatar 2012 OSAC Crime and Safety Report, April 8, 2012*, US Department of State, Bureau of Diplomatic Security.

US Department of State (2012b), *Oman 2012 OSAC Crime and Safety Report, March 20, 2012*, US Department of State, Overseas Security Advisory Council (OSAC).

US Department of State (2012c), *Country Reports on Terrorism 2011*, US Department of State Publication, Bureau of Counterterrorism, April, see http://www.state.gov/documents/organization/195768.pdf.

Utilities ME (2012), Quenching the Middle East's thirst responsibly, *Utilities ME*, March 11.

Utilities ME (2013a), No plans to raise power, water rates: UAE minister, *Utilities ME*, September 5.

Utilities ME (2013b), Bahrain said to plan revamp of energy subsidies, *Utilities ME*, December 9.

Utilities ME (2014a), Qatar invites bids to build "mega-reservoirs", *Utilities ME*, April 20

Utilities ME (2014b), Nuclear gives $1.7bn boost to UAE companies, *Utilities ME*, April 28.

Utilities ME (2014c), ENEC installs condenser at Barakah plant, *Utilities ME*, February 19.

Vaez, A. and K. Sadjadpour (2013), *Iran's Nuclear Odyssey: Costs and Risks, Report*, Washington, DC, Carnegie Endowment for International Peace, April 2.

Varghese, J. (2013), Daily per capita water usage in Qatar: 500 liters, *Gulf Times (Qatar)*, November 26.

Veiga, A. (2011), Food defence and security: the new reality, in Alpas, H., S. M. M. Berkowicz, and I. Ermakova (eds.), *Environmental Security and Ecoterrorism*, Springer, NATO Science for Peace and Security Series C: Environmental Security.

Vetrovec, S. (2004), Migrant transnationalism and modes of transformation, *International Migration Review*, 38, pp. 41–65.

Vickers, A. (2001), *Handbook of Water Use and Conservation*. Amherst, MA, Waterplow Press.

Vidino, L. (2013), The Arab Spring and the Middle East's monarchies, *International Relations and Security Network (ISN)*, March 12.

Vos, J., R. Boelens, and P. Mena (2014), From local to virtual water control: the globalization of water insecurity and water access conflicts, *Global Water Forum*, May 13.

Vörösmarty, C. J., P. B. McIntyre, M. O. Gessner, *et al.* (2010), Global threats to human water security and river biodiversity, *Nature*, 467, pp. 555–561.

Voutchkov, N. (2012), *Desalination Engineering: Planning and Design*. New York, NY, McGraw-Hill Professional.

Walker, B. H. and D. Salt, (2006), *Resilience Thinking*, Washington, DC, Island Press.

Walz, J. D. (2010), Islamic foundations for effective water management: four case studies, Masters' Thesis, Austin, TX, The University of Texas at Austin.

WAM (2011), Mohammed doubles free water quota for nationals, *Emirates*247, October 6.

WAM (2013a), Abu Dhabi targets zero water wastage in 1 year, *Emirates News Agency (WAM)*, March 27.

WAM (2013b), Abu Dhabi on groundwater tapping alert, *Emirates News Agency (WAM)*, March 13.

Ward, F. A. and Michelsen, A. (2002), The economic value of water in agriculture: concepts and policy applications, *Water Policy*, 4, pp. 423–446.

Warner, J. (2013), The Toshka mirage in the Egyptian desert: river diversion as political diversion. *Environmental Science and Policy*, 30, pp. 102–112.

Watani.ae. (n. d.), In Arabic, see http://www.watani.ae/portal/ar/home.aspx.

Waughray, D. (ed). (2011), *Water Security: The Water–Food–Energy–Climate Nexus*, London and Washington, World Economic Forum Water Initiative and Island Press.

WEF (2009), *Charting Our Water Future: Economic frameworks to inform decision-making*. Washington, DC, Island Press, World Economic Forum.

WEF (2011), *Water Security: The Water–Food–Energy–Climate Nexus*, Washington, DC, Island Press, World Economic Forum.

Weick, K. E. (2004), Normal accident theory as frame, link, and provocation, *Organization and Environment*, 17, pp. 27–31.

Wessendorf, S. (2008), Culturalist discourses on inclusion and exclusion: the Swiss citizenship debate, *Social Anthropology*, 16, pp.187–202.

Welton, G. (2011), *The Impact of Russia's 2010 Grain Export Ban*, Oxfam Research Reports, June 28, see http://www.oxfam.org/sites/www.oxfam.org/files/rr-impact-russias-grain-export-ban-280611-en.pdf.

Wesselink, E. and E. Weller (2006), Oil and violence in Sudan: drilling, poverty and death in Upper Nile State, *Multinational Monitor*, May/June, 27.

Whitlock, C. (2004), Consulate attack ends calm in Saudi Arabia, *Washington Post*, December 7.

Wickström, L. (2010), Islam and water: Islamic guiding principles on water management, in Luomi, M. (ed.), *Managing Blue Gold: New Perspectives on Water Security in the Levantine Middle East*, Helsinki, The Finnish Institute of International Affairs, FIIA Report, 25, pp. 98–109.

Wilkinson, S., T. Holdich, Mr. Amery, Mr. Hogarth, and H. J. Mackinder, (1904), The geographical pivot of history: discussion, *The Geographical Journal*, 23, pp. 437–444.

Williams, M. (2012), Future cyber attacks could rival 9–11, cripple US, warns Panetta, *IDG News Service, PC World*, October 12.

Wilner, A. (2011), *US and Iranian Strategic Competition: Iranian Views of How Iran's Asymmetric Warfare Developments Affect Competition with the US and the Gulf, September 2010–February 2011*, Washington, DC, Center for Strategic and International Studies.

Wimmer, A. and N. G. Schiller (2002), Methodological nationalism and beyond: nation-state building, migration and the social sciences, *Global Networks*, 2, pp. 301–334.

Woertz, E. (2013), *Oil for Food: The Global Food Crisis and the Middle East*, Oxford, Oxford University Press.

World Bank (1988), *Sultanate of Oman: Recent Economic Developments and Prospects. Report No 6899-OM*. Washington, DC, World Bank.

World Bank (2004a), *Seawater and Brackish Water Desalination in the Middle East, North Africa and Central Asia*, Washington, DC, World Bank

World Bank (2004b), *Water Resources Sector Strategy: Strategic Directions for World Bank Engagement*, Washington, DC, World Bank.

World Bank (2012), *Renewable Energy Desalination: An Emerging Solution to Close the Water Gap in the Middle East and North Africa*, Washington, DC, World Bank.

World Bank Data (n.d.), Average precipitation in depth (mm per year), see http://data.worldbank.org/indicator/AG.LND.PRCP.MM.

World Bank Data (n. d.), World Bank, see http://search.worldbank.org/data?qterm=urbanandlanguage=EN.

World Economic Forum (2014), *The Global Risks 2014*, ninth edition. Cologny/Geneva, World Economic Forum.

World Nuclear Association (2014), *Nuclear Desalination*, London, World Nuclear Association.

Worth, R. F. (2014), Leftward shift by conservative cleric leaves Saudis perplexed, *New York Times*, April 4.

WWAP (2003), *Water for People, Water for Life. World Water Assessment Programme, UN World Water Development Report*. Paris, United Nations Educational, Scientific and Cultural Organization (UNESCO).

WWAP (2009), *Water in a Changing World, The Third World Water Development Report, World Water Assessment Programme*, Paris, The United Nations Educational, Scientific and Cultural Organization (UNESCO).

WWAP (2012), *The United Nations World Water Development Report 4: Managing Water under Uncertainty and Risk, World Water Assessment Programme*, Paris, The United Nations Educational, Scientific and Cultural Organization (UNESCO).

Yan, D. and B. Chang (2011), Agricultural group seeks more overseas expansion, *China Daily*, March 14, p. 6.

Zeigler, L. (2013), A kingdom's thirst: the Saudi water challenge, *SUSTG*, March 30.

Zeitoun, M. (2009). All is not quiet on the waterfront, *Seed Magazine*, May 14.

Zerovec, M. and M. Bontenbal (2011), Labor nationalization policies in Oman: implications for Omani and migrant women workers, *Asian and Pacific Migration Journal*, 20, pp. 365–387.

Zwemer, S. M. (1907), Oman and Eastern Arabia, *Bulletin of the American Geographical Society*, 39, pp. 597–606.

Index

Lightning Source UK Ltd.
Milton Keynes UK
UKOW07n2238110815

256776UK00004B/92/P